Central Highland Redlegs

A Year in the Vietnam Central Highlands and Tri-Border Area

(September 1967 – September 1968)

BY

Robert W. Barrett

Stephen T. Stark

The military term *Redlegs* dates to the mid-1800s when the United States Army decided to use different colors to identify the various combat branches and chose Saxony blue for infantry, yellow for cavalry, and scarlet for artillery. During the Civil War, artillerymen wore a wide strip of scarlet cloth sewn vertically along the outside legs of their uniform trousers and the term *Red*leg was born. Throughout the Vietnam War, *Redleg* was used to refer to field artillery in general while the men serving as artillerymen were called *Redlegs*.

The Vietnam Central Highlands was a largely mountainous and jungle-covered region of South Vietnam that included generally the provinces of Kontum, Pleiku, Phu Bon, Dar Lak and Quang Duc where the First Brigade of the 4th Infantry Division served with distinction during some of the heaviest combat of the Vietnam War from 1966 until 1970.

The Tri-Border Area is a general region of the Central Highlands located in the vicinity of the confluence of the borders of Laos, Cambodia, and Vietnam.

Redlegs of Bravo Battery, 6th Battalion, 29th Artillery, pose for a quick snapshot on a firebase in the Central Highlands of South Vietnam west of Dak To in 1968. Photo provided by Nick Gillen courtesy of Bob Barrett who previously commanded Bravo Battery.

This book is co-authored by Robert W, Barrett and Stephen T. Stark. We agreed to write our personal accounts of our year in Vietnam with the 6th Battalion, 29th Artillery, 4th Infantry Division individually, then combine them into one book. Many others contributed to this book and will be recognized in the Dedication.

Central Highland Redlegs

A year in the Vietnam Central Highlands and Tri-Border Area
(September 1967 – September 1968)

by

Robert W. Barrett
Stephen T. Stark

Printed in the United States of America

ISBN 978-1-951921-05-7

Contents

Part One: Stephen Stark

Part Two: Robert Barrett

Foreword

Robert W. (Bob) Barrett (Captain, later Lieutenant Colonel), and Stephen T. (Steve) Stark (Private First Class, later Specialist 4) served as artillerymen with the First Brigade of the 4th Infantry Division in the Tri-Border Area of the South Vietnam Central Highlands from September 1967 until September 1968 during a time of great conflict as many of America's finest sons pursued, located, fought and defeated a numerically superior enemy during every major engagement and repeatedly drove him from the field of battle back west across the South Vietnam border into Cambodia and Laos at a time when we were forbidden to follow them.

Today, more than a half-century later, we clearly recall tiny details of some witnessed events while other memories are just shadowy images, snapshots in time, that sometimes come at night in the form of dreams. Like many who served, recollections of our tour of duty seems sometimes to run together as one long string of events that we attempt here to organize based on our firsthand experiences and with help from some of the men with whom we served, many graciously sharing their memories during reunions and other gatherings, and through telephone conversations and in personal correspondence.

During May of 2013, we visited the National Archives and Record Administration (NARA) located just outside Washington D.C. at College Park, Maryland to see what the official records had to say about the 6th Battalion, 29th Artillery in Vietnam during 1967 and '68. When we first arrived, they checked our identification and processed us per NARA procedures, then we met with NARA officials to explain what we were looking for. Amazingly, in just under four hours, they presented us with a hand-cart containing six storage boxes full of documents and escorted us to a large and well-lit open room with many tables where researchers and professional writers sat quietly shuffling through papers. They assigned us a table and we opened the first box only to instantly realize the tremendous volume of information before us and immediately decided to quickly preview each document and then scan only the ones of interest.[1] During that process, each individual piece of paper had to be handled separately and individually, then placed back into the box before removing another as required by the strict NARA protocol – they watched everybody in the room like a hawk.

The following morning, they delivered another hand-cart this time with eight boxes. We eventually scanned more than eight hundred pages including field notes, after-action reports, daily logs, summaries, hand-drawn maps, commanders' lessons learned reports, and other documents – all originals. We were amazed at the large volume of information available which helped us better

---

[1]*Several weeks before our arrival we requested and received NARA approval to bring a small portable scanner and laptop computer for copying documents. Familiarity with their rules and procedures before arriving helped smooth the process considerably.*

understand the order in which certain events occurred while also revealing a tremendous amount of additional information, much of which confirmed our personal recollections while some did not.

Regarding many of the field reports, it is important to remember that company-grade officers and others had strict orders to first get the job done and then – only then – write about it. The drudgery of report writing was a never-ending but necessary chore that was occasionally done daily but more commonly later on; maybe a day or two but sometimes a week or more when details were put to paper from scratch notes or sometimes from memory. After-action reports and other field memoranda were often hand-written and then transported by helicopter or convoy to another location further to the rear to be typed on manual typewriters, distributed per Army protocol, sent again further to the rear for review by others, then read again, over and over, as details were gathered for yet other reports even higher up the chain of command. Eventually, most of the original information thought to be of no further use was destroyed or sometimes sent stateside to one Army installation or another where it would again be picked over with much of it eventually destroyed. The surviving documents ultimately found their final resting place at NARA in Maryland, including many original unedited lower-echelon reports and notes that somehow survived, and even rough hand-drawn maps and sometimes notes scribbled on memo pads that we found during our visit there.

Robert W. (Bob) Barrett, (LTC Ret.) with hand-cart containing NARA files 13 May 2013

It is important to remember that the 1960s was a time before cellular telephones, digital scanners, or microprocessor-based computers like desktops and laptops in wide use today. Digital cameras did not exist – we used film that had to be processed and the only way to share photos was through the mail. Real-time communication in Vietnam was in person, by telephone or by radio.[2] Especially compared to today, information moved at a snail's pace.

Some of the hand-carried field reports, memos, orders, and other documents never made it to the rear because they were misplaced, lost, destroyed or perhaps even misdirected through the maze of the U. S. Army recordkeeping systems – all done manually. This, we think, helps explain some of the holes in the official written history – the missing pieces in the official Army records we found at NARA regarding events that we remember so clearly or documented in our personal notes, letters home, photographs, and in other records.

We realize that some of the witnessed events that seemed so terribly important to us when they happened were actually so small and insignificant that they were not mentioned in reports or memos at any level.

Also, like all wars, some events were classified for one reason or another and intentionally never mentioned. Even many of the documents we viewed at NARA were still marked as classified and had to undergo on-site reclassification process by the ever-present NARA staff personnel before we were allowed to scan them.

Off and on during his tour of duty, Steve Stark served a total of roughly five months as an Artillery Liaison Specialist assigned to the First Brigade, stationed either at Dak To or on other assignments to the various fire support bases and other locations within our area of operation. During the Battle of Dak To[3] which lasted most of November 1967, and during Tet[4] beginning very early in February 1968, he was often at the Frist Brigade TOC fourteen to sixteen hours a day and was witness to the planning, preparations, and execution of many combat operations against the NVA and VC in the Central Highlands.[5] When time allowed, he kept a personal notebook (a diary of sorts) that included many significant events, dates, times, detailed notes, and personal observations that has also been used as reference material.

Some of the information presented here is gleaned from the many photographs we took, or that have been provided by others over the years which have sometimes revealed subtle details about

---

[2] *Even telephone calls home to check on newborn babies or sick relatives were not possible except at a few places in Vietnam and then only through MARS, the acronym for "Military Auxiliary (today Affiliate) Radio System" using short-wave radios operated primarily by civilian ham radio operators and other volunteers. Nobody I knew while in country was able to call home during his tour of duty except during their R&R.*

[3] *Although commonly called the "Battle of Dak To," Battle for Dak To" is the terminology used by our senior 4th Infantry Division leaders and others in reports and correspondence. This difference, however, is insignificant and is mentioned here only for historical interest.*

[4] *Tet (sometimes TET) is the Vietnamese lunar new year celebrated throughout the country.*

[5] *NVA is an abbreviation for North Vietnamese Army. VC stands for Viet Cong.*

people, places, and events that might otherwise have been lost. Where photographs are shown, the original photographer, if known, is credited. When the photographer is not known, we have identified the source of the photograph.

Yet another valuable source of excellent background information came from a small memoranda book carried daily by Major Leslie D. Crouse beginning in January 1968 when he arrived in country until he was killed in action on 31 August that same year. Major Crouse's son, Dewayne O. "Doc" Crouse, traveled to Branson, Missouri from his home in Alaska for a reunion of the 6th Battalion, 29th Artillery over Veterans Day week in November 2014 and presented his father's memo book to us. We later studied and transcribed all 42 pages of the dates, notes, names, reminders, radio frequencies, map coordinates, sketches, plans, scribbles and other contents for additional clues to our history and presented our findings to the Crouse family in 2017. Additionally, Doc Crouse also shared parts of his father's letters home to Mrs. Crouse in 1968 that include personal and private insight into the mind of a good and decent man who is still missed greatly by his family and by those of us who served with him. We thank the Crouse family for their kindness in permitting us to share parts of Major Crouse's notes and letters.

Photo of the small memoranda notebook carried by Major Leslie Crouse laying on map of Dak To area next to U. S. Army lensatic compass.

Additionally, the personal memories of some of the men with whom we served has been invaluable in piecing things together. Their accounts, some written in their own words[6], are included separately and under their respective names with their permission or, if they are now deceased, with permission of their family.

[6] *We have intentionally published their accounts and writings exactly as provided to us with no modifications and in a special section near the end of this book. Where the author is deceased, his family has given their permission.*

Information obtained from other sources is listed in the bibliography and we have also provided a reference section that contains some of the commonly used acronyms along with unofficial definitions. We have also included a small amount of technical details that will hopefully be of interest to some readers.

We are grateful to the soldiers, authors, historians and others who have contributed to our work by providing details to help us tell our history.

We need to say here that, sadly, much of the available information on the Vietnam War is somewhat like the stuff regularly produced by horses and cows and not fit for human consumption. We believe that a healthy dose of skepticism is in order when reading any account of the Vietnam War, whether a book or magazine article or on the Internet; or while watching movies or television programs. Even some of the so-called "history" is seriously flawed, being politically biased or, worse yet, intentionally deceptive for one reason or another. Our hope is that people interested in the truth about our war will take the time to study the available documentation for themselves and most importantly always ask the question, "*What are they trying to sell me...and why?*"

No historical account is perfect, and mistakes have undoubtedly been made for which we take full responsibility and ask the indulgence of the reader.

Our primary hope is that the families, loved ones and friends of the men with whom we served will gain something of value in reading this history that includes individual accounts of some of the men with whom we served.

When we first began writing, we planned to combine our individual recollections into a single presentation but soon realized that writing separately would make more sense because we had significantly different assignments and were usually at different locations. Therefore, we have divided this book into two separate sections under each of our names although we often address similar events in different ways as we knew them.

The following sections are written by Steve Stark and include personal recollections and some of our common history.

Dedication

My part of our book is dedicated to the fine men of the 6th Battalion, 29th Artillery with whom I served and the families, friends and loved ones of those killed during combat operations in pursuit of the North Vietnamese and Viet Cong in the Central Highlands of the Republic of South Vietnam.

My wife of almost fifty years, Katie Ann Stark, gets special thanks for encouraging me to continue writing even as I grew weary of it. She also occasionally helped gently pry some of the more shadowy details from my mind in a way that only a best friend can do.

My Captain in Vietnam, now my faithful friend, Robert W. (Bob) Barrett, later Lieutenant Colonel United States Army, continues earning my respect even after more than fifty years since our time together in the Central Highlands where he served during the Battle of Dak To with the 5th Battalion, 16th Artillery as an artillery Liaison Officer, and later as Commanding Officer of two batteries of the 6th Battalion, 29th Artillery – Bravo Battery and Headquarters Battery. A professional soldier in every way, Bob is an excellent and expert artilleryman who honed his many skills over decades of faithful and dedicated service to the United States Army and to our nation in other ways after his military retirement. His unending patience in helping me better understand many of the technicalities and logistical details of some of the events we witnessed is greatly appreciated, as is his friendship.

The Beginning

After a couple of years of college that were taking me nowhere, I volunteered for the draft in late 1966 and began Basic Training at Fort Leonard Wood, Missouri on 24 April 1967, graduating on 22 June exactly sixty days after taking the oath of enlistment. [7]

Just before the graduation ceremony and parade, they handed out orders which would scatter most of us to Army posts throughout the United States for Advanced Individual Training (AIT) with the majority becoming infantrymen while others would be medics, mechanics, truck drivers, clerks, military policemen, radio operators, and to learn any of hundreds of other military jobs – called an MOS (Military Occupation Specialty). Although most would begin their time of service as enlisted soldiers, others would receive training to become officers.[8] I was ordered to report to Fort Sill in southwestern Oklahoma to study field artillery at the United States Army Field Artillery Training Center (USATC-FA).

My Dad and Mom drove the 140 miles to Fort Leonard Wood for the graduation ceremony and afterwards I traveled home with them to Webster Groves for three days spent with family and a dwindling number of friends, some hiding from the draft.

On the morning of 26 June 1967, I boarded a commercial flight from Lambert Field, St. Louis, and landed at Lawton, Oklahoma near Fort Sill early-afternoon on a beautiful sunlit day with mild temperatures and a light, dry southerly wind. I liked Fort Sill and Oklahoma from the very beginning but had no idea that the Sooner State would eventually become my home and the birthplace of our three children and five of our eight grandchildren.

At historic Fort Sill we billeted in one of the old, stone, three-story barracks still standing on Randolph Road about a half-mile west of post headquarters at McNair Hall. We were in easy

---

[7] *I trained in Company C, 2nd Battalion, 2nd Basic Training Brigade, 5th Army and our Company took top honors which put us first in the graduation parade which was a big deal to many of us at the time. Most of us were draftees who had taken an oath to serve for two years and our service numbers began with the letters "US," then followed by eight digits. Those who were not drafted but instead signed up for four-year enlistments had service numbers that began with "RA" that stood for Regular Army. Also, there were a few in our unit whose service numbers began with "ER" (Enlisted Reserve) or "NG" (National Guard.). Throughout my two-year enlistment I never noticed any difference in the attitude, performance or esprit de corps of anyone, regardless of his service number prefix.*

[8] *Some in our unit had qualified for Officers Candidate School (OCS) or Warrant Officer Flight School but only a few volunteered due to the additional required military commitment. I turned down both offers because I did not plan on making the Army a career.*

walking distance of the old guardhouse where the great Chiricahua Apache Geronimo was held prisoner and often incorrectly reported as having worn a deep rut in the rock floor.[9]

Our AIT class at Ft. Sill was ASSC 30 (Artillery Survey Specialists Course 30) and study began the next day when we were called to formation in the large parade field behind the buildings. Training there was fast-paced and gave us precious little time for much anything but long classroom sessions, a lot of field time, and our nightly studies. We were told that if sent to Vietnam, which was probable, that we would be doing a lot more than just surveying-in artillery pieces and they were teaching us other valuable skills.

Our instructors taught us artillery history beginning with the U. S. fight for independence against the British and ending with the Korean War. The course of study included map reading, distance determination, position determination, targeting, ground reconnaissance, and the correct use of magnetic compasses. Later we learned the proper use of ground survey equipment including rod and tape, aiming circles, T-2 and T-16 theodolites. We studied some algebra but primarily geometry and learned to do triangulation problems almost in our sleep. On the hot and windy ranges at Ft. Sill we practiced what we had learned in the classroom which concentrated on altimetry, astronomic observations (navigating using the sun, moon, and stars), and coordinate conversions – all without electronic calculators or computers. Everything was done manually with pencil and paper and using graphs and tables.

Our field survey work relied primarily on using the well-known spot elevations of the block houses positioned atop the many hills around Fort Sill's firing ranges and measuring long distances using the identical rod and tape methods employed for centuries. We worked the longer-distance problems using heavy and awkward sound-ranging equipment that did not always work and took an hour to set up and another half-hour to calculate a distance. They taught us many practical rule-of-thumb angle and distance estimating methods but always emphasized the importance of precise measurements of time, distance, angles and elevation. During our course of study, we also spent about a day learning generally about how a fire direction center (FDC) works and received a crash-course in the new FADAC[10] computers.

During the last few days of training, our instructors reviewed what we had learned by verbally quizzing us first as a group and then individually. They asked questions and listened carefully to our answers and allowed us to elaborate while they took notes. I do not recall a final examination, but we had already taken many tests and quizzes throughout the training process during which a

---

[9] *Geronimo was held behind bars only a short time and later roaming freely, taking up farming nearby until his death on 17 February 1909. He is buried not too far from the Old Post where later a small pyramid of round stones was constructed and where nearby graves hold the remains of other American warriors.*

[10] *FADAC is an acronym for "Field Artillery Digital Analog Computer."*

very few, perhaps two or three, had washed out and were sent elsewhere to learn a different Army MOS. Artillery survey at that time required the application of mathematics and physics and was not suited for everyone. I was never a star student or whiz at math, but from an early age I had a good mind for the physical sciences and understood angles and maps and how to use a compass thanks to my Dad and other great boyhood teachers.

When we fell into formation on the morning of 22 August 1967, the Captain standing before us said, "First, our war heroes" and called the names of about a dozen of us including Glen Farasy, Mike Alesi, and me and told us to step forward. He handed each of us our orders for Vietnam and we stepped back into formation. The remainder of our classmates were also given orders but for other places around the United States and overseas. Many of them later would also serve in Vietnam.

We spent two days learning about the M16 rifle,[11] first during a half-day of familiarization, disassembly, cleaning and reassembly and then a day and a half at the firing range working first close-in and then at more distant targets. To me, the relatively new so-called *black plastic gun* was much easier to fire than the M14 that I had carried during basic training thanks primarily to its weight and much lower recoil and I scored at the top of our class and earned an expert rating. Although much difference in appearance and operation, to me the M16 had a certain familiarity because it shot sort of like my Dad's .30 Carbine[12] he had taught me to handle "with proficiency" as he put it.

On 23 August 1967, Class Number 30-67 of the Artillery Survey[13] Specialist Course (ASSC-30) graduated following 59 days of rigorous field and classroom training.[14]

The following two days (24 and 25 August), around a hundred recent graduates of various artillery schools boarded trucks to an outdoor training area with open-sided tents for what they called *RVN*

---

[11] *During basic training at Fort Leonard Wood we learned to handle, fire and maintain the M14 rifle which chambered a 7.62x51mm round (30 caliber, 7.62 NATO) that had been adapted by the Army in the 1950s after the Korean War. Now in 1967, the M16 was the Army's choice because the experts decided it was better suited for close-in jungle warfare than the M14.*

[12] *As an artillery officer, my Dad had carried an M1 Carbine during WWII and purchased one like it through a CMP program supported by the National Rifle Association for around $20.00 (shipping included) in the early 1960s.*

[13] *The Army Field Manual 6-2 (FM 6-2) states: "The mission of artillery survey is to provide a common grid which will permit the massing of fires, the delivery of surprise observed fires, the delivery of effective unobserved fires, and the transmission of enemy location and target information from one unit to another."*

[14] *Our class was small and consisted of about two dozen including a U. S. Marine named Gunny Hannah who sometimes marched us to class counting cadence the Marine Corps way. He also sometimes drank beer with us in Lawton. Semper Fi, Gunny, wherever you are.*

Orientation Training that was required for all replacements headed for Vietnam. They covered many subjects including what to do if we were captured, also reading recent reports about the cruel treatment that American POWs were enduring at the hands of the North Vietnamese Communists. They reminded us about the U. S. military Code of Conduct everybody memorized during basic training but now it suddenly had special meaning. They made sure we clearly understood the many horrors associated with being taken prisoner including details of some of the more gruesome torture methods employed by the NVA. To me, the message was crystal clear: Don't ever let them capture you.

I returned home by bus to Webster Groves to spend about three weeks with family and friends and to say goodbye. On 14 September 1967 I boarded a commercial jet at Lambert Field in St. Louis and flew to California, ending up at Travis Air Force Base just north of Oakland.

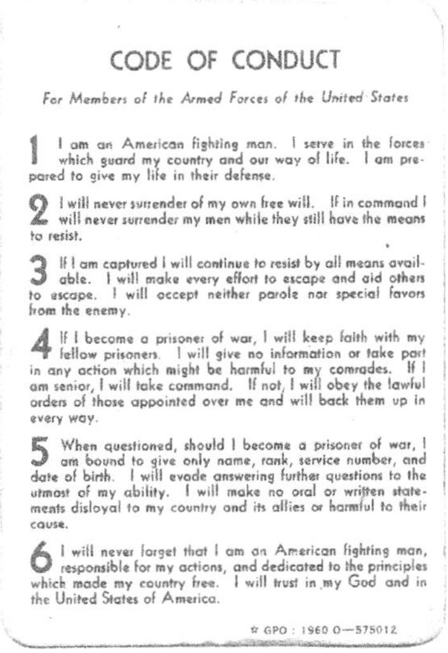

CODE OF CONDUCT

For Members of the Armed Forces of the United States

1 I am an American fighting man. I serve in the forces which guard my country and our way of life. I am prepared to give my life in their defense.

2 I will never surrender of my own free will. If in command I will never surrender my men while they still have the means to resist.

3 If I am captured I will continue to resist by all means available. I will make every effort to escape and aid others to escape. I will accept neither parole nor special favors from the enemy.

4 If I become a prisoner of war, I will keep faith with my fellow prisoners. I will give no information or take part in any action which might be harmful to my comrades. If I am senior, I will take command. If not, I will obey the lawful orders of those appointed over me and will back them up in every way.

5 When questioned, should I become a prisoner of war, I am bound to give only name, rank, service number, and date of birth. I will evade answering further questions to the utmost of my ability. I will make no oral or written statements disloyal to my country and its allies or harmful to their cause.

6 I will never forget that I am an American fighting man, responsible for my actions, and dedicated to the principles which made my country free. I will trust in my God and in the United States of America.

☆ GPO : 1960 O—575012

Code of Conduct printed on 2½ by 3½ - inch card and carried by American soldiers.

"Welcome to Disneyland in Green"

Late afternoon of 18 September 1967 they marched us out onto the tarmac at Travis in a lose formation with shoulder-slung duffle bags and wearing our low-quarter shoes, khaki trousers, short sleeve shirts, and garrison caps. Some of us also carried Dopp kits or small so-called AWOL[15] bags holding personal items, mostly magazines and other reading material. Almost everybody held the rank of Private or Specialist, and as we got closer to the TWA jetliner each tossed his duffle bag onto one of the large airport carts and then walked up the portable stairway where two stewardesses[16] stood just inside the plane to welcome each man with a warm smile. We made our way inside to fill the plane from the rear to the front with no assigned seats – get in, sit down, buckle up. My seat was near the front of the plane on the left side and I was glad to get a window seat. The jetliner was packed completely full and nobody talked as we taxied out and finally lifted off into the California sky for a very long journey into the unknown.

The stewardesses prepared light meals during the flight, and we landed for fuel at Anchorage and again several hours later in Japan. Both times we were allowed to get off and stretch our legs and get a cup of coffee and a snack. On both occasions I also mailed home post cards to my folks.

Everybody slept off and on during the long and tiring flight and we finally arrived over Cam Ranh Bay just south of Nha Trang on the southeast coast of South Vietnam on 20 September 1967. As we circled to land, we looked down on tiny fishing boats on the water near the shoreline and spotted dozens of small wooden-frame buildings, some with grass thatch roofs and others of tin.

The plane taxied near a large hangar and we unceremoniously disembarked while U. S. Air Force ground personnel unloaded and tossed our duffle bags onto the tarmac. I do not remember seeing the stewardesses as we deplaned, but I have been told that some of them had a difficult time saying goodbye. Bob Barrett clearly remembers seeing a stewardess with tears in her eyes as he arrived in country at about that same time in September. After collecting our bags, we boarded busses for a short ride to brand new two-story wooden barracks where on the second floor I selected a bunk and unpacked part of my duffel, then cleaned up and went downstairs where I spotted Glen Farasy – it was good to see a familiar face. He was the only one I recognized at Cam Ranh Bay.

Glen and I had grown up in the same town, graduated from the same high school, were sworn in at the exact same time at the St. Louis enlistment center, and stood in line together when they made our dog tags and handed out our eight-digit service numbers – ours are identical except for the last

---

[15] *A small carrying bag often purchased at the PX used to carry personal possessions while soldiers were on leave.*

[16] *Flight attendants were properly called "Stewardesses" back then and were all lovely ladies of impeccable appearance and excellent manners. They took great pride in seeing that all passengers on every flight were treated kindly, and especially GIs on the way to war.*

digit. We were assigned to the same basic training brigade at Fort Leonard Wood and then both selected for Artillery Survey training ending up in the same small Advanced Individual Training class at Fort Sill. Now just five short months after taking the oath of enlistment we were in the same exact place in Vietnam. It's a small world.

We had a bite to eat and then took off walking to explore the base. As we came across a group of Vietnamese civilians working near the mess hall, Glen exclaimed, "Hey Steve, look at all the foreigners!" which is exactly the kind of ironic understatement for which he is now famous.

I was surprised at the large number of Vietnamese civilians working on the base, many wearing the familiar loose-fitting black pajama-like outfits and pointed *nón lá* (translated literally to "leaf hat"). The only other people I had ever seen dressed that way were on the evening news as American GIs dragged them from spider holes in the jungle or from caves. At that time, I thought our enemy were all civilians during the day and VC terrorists by night. It still amazes me that despite all the training received at Fort Leonard Wood and Fort Sill, I still had no idea that we would be going up against a well-trained and well-equipped army of NVA soldiers – definitely not what Walter Cronkite was telling the American people on the evening news.

The following day we continued acclimating to the sweltering environment and overcame our jet lag. We had no specific orders yet, or duties – we just sort of hung around. The mess hall was more like a restaurant than anything I'd seen at Fort Leonard Wood or Fort Sill with a wide variety of great food including hamburgers and shakes. I remember thinking that maybe Vietnam wasn't so bad after all. Cam Ranh Bay was very peaceful, and I slept very well while there.

They awakened us each morning around sunup and we made our bunks, swept the barracks, showered, shaved, and dressed in the uniforms we had brought with us – stateside fatigues, combat boots, and OD baseball caps. We received in-country orientation and they handed out Geneva Convention and other wallet-sized cards printed on heavy paper. Everything was done in a precise and specified way and always in groups of roughly fifty men – it was an assembly line. After breakfast each morning we were called to formation as they read the names of new arrivals and then handed-out travel orders for those headed to various locations throughout South Vietnam. The third evening after chow they called us together to sign more paperwork including a declaration of our next of kin which was a sobering experience. At that time, the Army paid ten-thousand dollars to beneficiaries of soldiers killed in Vietnam.
During the morning formation on the fourth day, many of us received travel orders to report to the 4th Infantry Division headquarters at Camp Enari[17] in the Central Highlands just north of Pleiku

---

[17] *Camp Enari was named in honor of First Lieutenant Mark N. Enari who during a fierce firefight on 2 December 1966, single-handedly charged an enemy machinegun nest to successfully rescue five wounded and trapped men in his command, posthumously earning the Silver Star.*

about 175 miles north of Cam Ranh Bay. We were given just fifteen minutes to gather our gear and scramble aboard the waiting busses for the brief ride to a C-130 waiting on the runway with engines already running. We walked up the ramp at the rear of the plane through the exhaust fumes and into the belly of the beast, a large empty hull with no seats.

The obviously well experienced Air Force Loadmasters[18] quickly and efficiently stacked us in like sardines all the way from the front to the rear of the cargo area. We sat side by side on the floor facing forward in rows, each separated by a heavy cargo strap running side to side and secured somehow to the inside of the plane. The aircraft flooring consisted of several large steel plates with a wide gap between each one. After everyone was boarded and sitting with our feet forward and the straps across our laps, one of the Loadmasters loudly announced over the engine noise that the steel plates would probably slide around a bit, especially during takeoff and landing, and that we should be careful about sitting on the cracks or we could lose part of our gluteus maximums or worse yet our *family jewels*. After the announcement and following the usual vulgar chatter, most everybody looked down between his legs as the plane began to taxi. To my knowledge, nobody was wounded on that trip.

The flight to Pleiku was loud and turbulent and the steel floor plates offered no comfort. We landed about an hour later and when the plane stopped the ramp lowered and the Loadmasters ordered us to stand up. We quickly exited down the ramp and tossed our duffel bags onto a waiting truck, then scrambled aboard one of several small Army busses waiting there. All the bus windows were wide open but covered with wire mesh and as we sped off, some clown in the back of our bus jokingly yelled out that the wire screen was to keep the Vietnamese girls away from us. Our driver, having probably heard that same comment a hundred times before, said one word: "grenades." Everything got quiet.[19]

---

[18] *A Loadmaster is a crew member in charge of aerial cargos including all people and equipment.*
[19] *The Viet Cong had a nasty habit of tossing hand grenades and satchel charges into vehicles, thus the wire screens.*

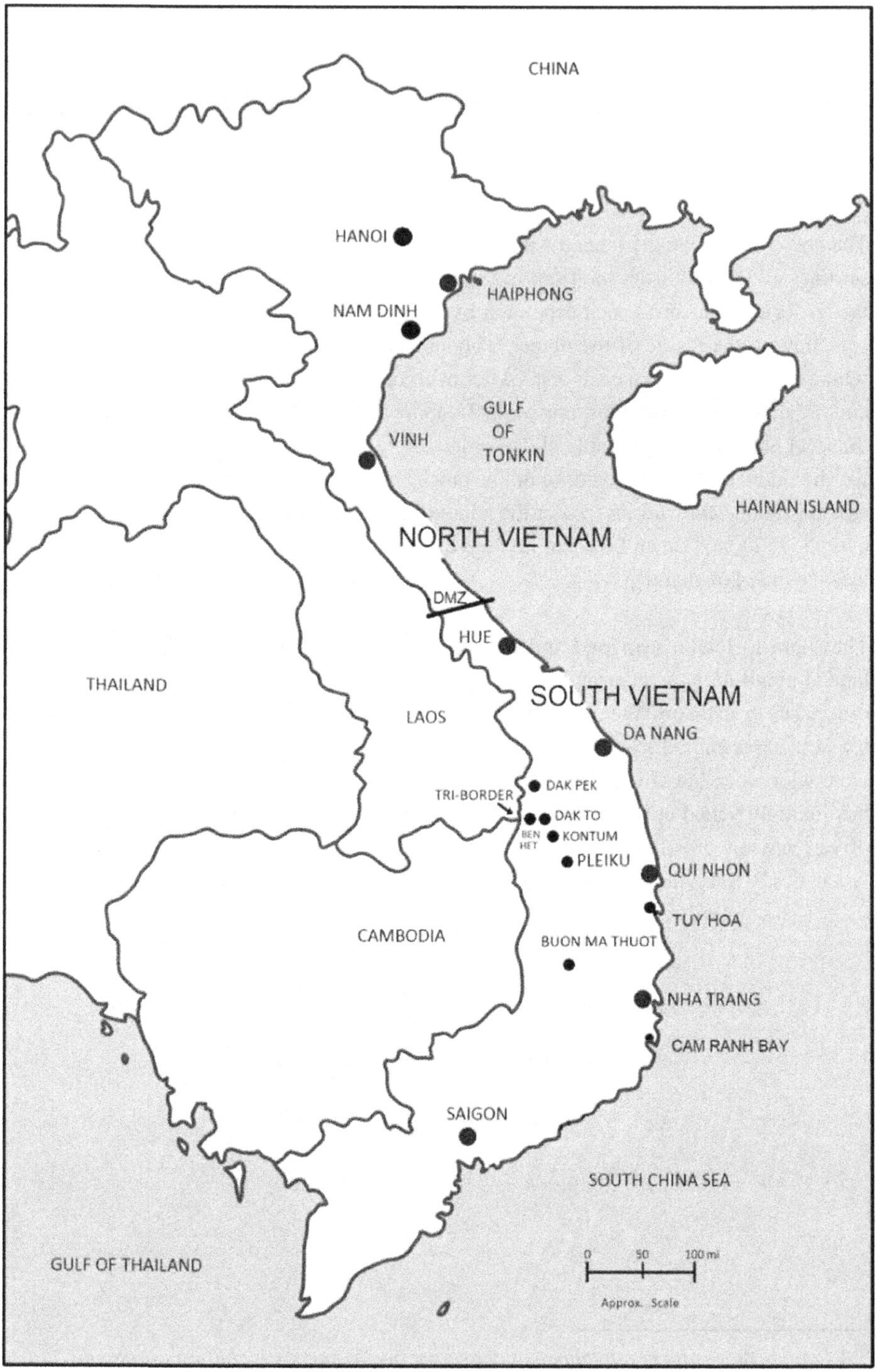

Map of Vietnam 1967- 1968

Camp Enari

Arriving at Camp Enari, we received a quick and very informal welcome from a 4th Infantry Division Officer and then an NCO. This, I think, is where I first heard the words, "Welcome to Disneyland in Green," but do not remember who said it.

We were billeted in GP Medium tents on wooden frames about a foot off the ground. Each tent was packed with Army metal bunks with a narrow corridor down the middle. They handed out mosquito nets and well-worn but clean Army sleeping bags. I slept very soundly my first night in the Vietnam Central Highlands.

Our mix of new arrivals had been trained in various military occupation specialties, most infantry and artillery; we were all first timers in Vietnam. The 4th Infantry Division had first arrived in country by ship just ten months earlier as a unit but now we were the replacements arriving in smaller groups or as individuals to replace the guys who would soon be headed home.

On the second day, we received additional in-country orientation which included instructions on taking malaria pills, personal sanitation, the importance of dry feet, the use of water purification tablets, basic Vietnamese history, in-country etiquette, and a very logical and well-delivered lecture on the many reasons we were fighting in South Vietnam which went something like – "…to uphold the promise of the United States of America to protect the South Vietnamese people from the invading North Vietnamese Communists and their allies including the Viet Cong, Chinese and others who would enslave their people…" It was clear to me that the officers who delivered the message believed every word of it and so did I. For the remainder of my entire time in country I do not ever remember anyone mentioning right or wrong regarding the war or hearing much anything about the complex politics or the discontent at home, mostly on college campuses and as always hyped by the news media. The only newspaper I read while there was the Stars and Stripes that was published by the Army and covered the protests only occasionally. My parents and friends who wrote never mentioned politics or anything about the war. I guess they figured I already had enough to worry about.

We were issued three pair of new jungle fatigues (to replace the stateside fatigues), green boxer-shorts and t-shirts, a web belt with a new kind of non-shiny GI buckle, black leather jungle boots with olive drab heavy canvas tops, helmet liner, steel helmet, two-sided helmet cover (green and brown camouflage), helmet cover band, pistol belt, butt pack (sometimes called a fanny pack), musette bag, four M16 ammo pouches, a plastic canteen, aluminum canteen cup, canteen cover, water purification tablets, one olive drab towel, a gas mask with pouch, a rubber poncho, a camouflage nylon poncho liner, a first aid pouch (with bandage), and a small plastic container of liquid insect repellant. We also were issued brand-new Colt M16 rifles and four 20-round metal magazines, a bayonet with scabbard, and a rifle cleaning kit and gun oil. But no ammunition.

On the third day we boarded deuce-and-a-half trucks for a trip to makeshift rifle range where they issued bandoleers of 5.56mm M16 ammunition. We broke down and thoroughly cleaned our rifles as the NCOs watched over us offering never-ending comments about one thing or another. Somebody warned us not to load our magazines with more than eighteen rounds – something every Vietnam veteran of that time seems to remember.[20] In groups of about fifteen, we first fired prone over sandbag rests at twenty-five-meters using the standard 1000-inch targets and adjusted our front and rear sights. Each of us then shot at more distant targets on semi-automatic and then full-automatic. Unlike previously at Fort Sill and Fort Leonard Wood, nobody kept score which, I suppose, is because everyone had already qualified stateside. Another thorough cleaning and inspection and we were finished for the day and trucked back to the compound for a hot meal served in a large mess tent.

On the morning of the fourth day, they formed us into five platoons of about forty men each and we marched in loose formations out of the compound and then each platoon broke off in different directions. Unlike basic training, the NCOs did not yell but simply walked among us and talked quietly as we moved along in irregular formations as they gave occasional instruction and offered reminders including watching for trip wires and punji traps. Every now and then one of them would say, "Would somebody *please* keep an eye on those damned trees?" There was, of course, the traditional and seemingly never-ending reminder to "spread out…don't bunch up." We were obviously still in training.

At one point, a few Montagnard[21] kids joined the march, happily walking along in groups of two or three, some wearing donated lose-fitting Army fatigue shirts and always smiling and happy; glad to take a candy bar or can of C-rations as a gesture of good will from GIs who had plenty and usually smiled back.

A Montagnard youngster joins us during our first outing. (Steve Stark photo)

---

[20] *The origin of this tradition is somewhat mysterious, but most experts seem to agree that it had something to do with the magazine design rather than the design of the rifle.*

[21] *Montagnards are the native peoples of the Central Highlands of several different bands who were strong US Army allies. The name comes from a French word meaning literally "mountaineer."*

We rested around midday for my first C-ration meal, then received instruction from a Fifth Special Forces Sergeant and his helpers on the proper application of primacord for clearing small obstacles and C-4 for other purposes including blowing enemy bunkers and tunnels. When the instruction was over, they detonated a half-brick of C-4 set in a quickly dug hole which blasted a large column of dirt into the air.

They also gave us a crash course in the setup and placement of trip flares and Claymore mines, subjects that were covered only generally during my basic training but that now were somehow much more interesting. Some of the men with us had just completed stateside AIT training and earned their infantry MOS called *Eleven-Bravo* (11-B) and had already been trained in such skills but even they seemed to appreciate the refresher.

Late that afternoon after walking through lightly wooded and hilly terrain a total distance of perhaps three convoluted miles, twice crossing our previous trace, we stopped on a grassy knoll partially covered with short bushes and near a Montagnard village about a half-mile away. Everybody worked to improve our night defensive position by cleaning up the foxholes that were already there, also filling and stacking cloth sandbags at various places around the perimeter. Most of the holes were little more than shallow depressions with a few sandbags stacked around to increase their effective depth. Digging in the loose red clay was easy and it was obvious that other new arrivals had used that same location many times before based also on the partially buried trash and empty C-ration cans – part of the area looked like a trash heap. There was also a latrine area about forty meters downhill at the end of a well-walked trail. Under close supervision by the NCOs, we set trip flares and claymore mines, then enjoyed a tasty dinner of cold C-rations washed down with lukewarm water.

They called out the night perimeter picket schedule, the plan being for about ten to remain awake in two-hour shifts while others slept – or tried to sleep. My time on the perimeter began sometime around midnight and lasted at least six hours – or so it seemed. Cloudy, pitch black sky, no wind, no light, and no sound except the occasional turning or snorting of a nearby GI and the buzzing of mosquitoes. The bug repellant seemed to work well on mosquitoes but did little against the ants and other nibbling creatures that were our constant companions.

I tried, unsuccessfully, to convince myself that the Army would never send such green troops into a *real* combat zone and that this was all just more training – another make-believe exercise like at Fort Leonard Wood where we would all stand up afterwards and listen to a Sergeant or Lieutenant tell us what we had done right and complain about the many things done wrong. Even so, I could not help remembering the war movies I'd seen where some green soldier on guard duty never hears the bad guy with the bayonet in his teeth slithering silently through the grass. My fourth night in the Central Highlands was not nearly as pleasant as my first three.

The 6th Battalion, 29th Artillery

Early the next morning we arose, ate, buried our trash, emptied the sandbags and then walked the mile or so back into our tent camp and cleaned up. That afternoon most of us packed our gear and loaded onto trucks to be taken to our individual units at Camp Enari. My new home, along with about twenty others, was the 6th Battalion of the 29th Artillery Regiment that was part of the First Brigade of the well-celebrated 4th Infantry Division.

Unit Insignia of the 29th Artillery Regiment formed in Kansas in 1918. The Latin words Fidelis et Verus translates to Faithful and True. Our battalion motto was "Can Do."

When we arrived in the battalion area, an officer welcomed us as a group. We were billeted in one of the several identical single-story, one-room wooden framed barrack, each having clapboard walls, wooden tarped roofs, roughly finished concrete floors, and five shuttered and screened windows for ventilation. Inside each barracks were ten or twelve metal Army spring bunks arranged in the traditional barracks fashion, each with its own wooden footlocker painted Artillery Red. In front of each building were water-filled and covered 55-gallon drums and buckets for fire protection also painted Artillery Red. The perimeter of each building was surrounded by sandbags stacked about four feet high and all the barracks entrances faced inward toward the rectangular sort of *parade ground* area. At the immediate east end of the parade ground was a tall wooden structure (an archway of sorts) of a somewhat Asian design that held high the 29th Artillery crest in all its glory. Immediately below the arch was a single small evergreen tree and nearby was a small arched bridge flanked by two small concrete ponds connected under the bridge. All around the battalion area were neatly arranged walkways lined with used artillery powder cannisters placed there to spruce the place up. They were also painted red.

6th Battalion, 29th Artillery at Camp Enari looking west in 1968. (photo courtesy of Bob Barrett)

After getting settled in, they told us to open our duffels and remove our stateside fatigues, stateside boots, and civilian clothing for storage in a connex[22] located near the motor pool in the battalion area. They assured us that everything would be safe and sound until it was time to go home.

Our battalion was comprised mainly of young men in our late teens or very early twenties with a few NCOs and more senior officers in their early thirties, but the Lieutenants were all about the same age as most enlisted men. Like most wars, the Vietnam War was fought by young men.

Two of my buddies from AIT, Mike Alesi and Glen Farasy, were also there but I do not remember the order in which we each arrived. Our barracks was occupied by artillerymen trained in a wide variety of specialties, but most were cannoneers. Those of us holding the 82 Charlie MOS were told we were now the *Artillery Survey Section.*[23]

---

[22] *A connex (sometimes "connex box") is a steel container that was developed for shipping military equipment during WWII. During the Vietnam War, the connex was commonly used for storage and other purposes including field TOCs and FDCs, sometimes welded together to increase their size. Connex is an acronym derived from "container express."*

[23] *Like many other units in Vietnam, the 4th Infantry Division organizational breakdown was: Brigade, Battalion, Regiment, Battery, and Section. So, in Vietnam we were: First Brigade, 6th Battalion, 29th Artillery Regiment, Headquarters Battery[23], Survey Section. A Section, therefore, was like an infantry Platoon, both which fluctuated significantly in its size depending on many variables including specific tasks and available manpower.*

It soon became apparent that our barracks at Camp Enari was only temporary and that most of us would be sent elsewhere as we were needed. Meanwhile, we were assigned various camp duties over the next few days as other recently arrived troops joined us.

We burned latrine cans and purified piss tubes and performed other duties including pulling KP and helping frame a new wooden building to house even more troops on the way to join us during what would become the largest buildup of the Vietnam War.

The cooks in our battalion mess hall provided three warm meals a day with breakfast by far the best because it was predictable: good coffee, powdered eggs, potato patties, ham or bacon, and a choice of brown biscuits or Texas toast and jelly. In the chow line on my second or third day there, after they had shoveled the eggs and meat onto my plate, I spotted a large pan holding steaming hot biscuits just out of the oven and grabbed a couple. Sitting down to eat, I broke open the first biscuit to add some jelly and a wonderful aroma arose with the steam. I was delighted to see that the Army cooks had baked in little bits of crunchy nuts – just like Mom used to make! Closer inspection however revealed no nuts, just well-done weevils. In the Army you learn all sorts of interesting things.[24]

I blamed the mess hall for my first case of dysentery, but it could have just as easily been the malaria pill or maybe even the water which always tasted and smelled to me like a cross between bleach and old gym socks. So, off to the medic's hooch for a quick diagnosis and something to ease my discomfort. Like many who served in Vietnam, I was plagued with stomach and digestive problems off and on during my entire tour of duty.

The 6th Battalion, 29th Artillery consisted of three firing batteries, A (*Alpha*), B (*Bravo*), and C (*Charlie*). Each battery consisted of six, 105mm howitzers designated "M 101 A-1," but with only five howitzers in the field at a time with one of each battery held in reserve at base camp. Each battery provided what was called *direct support* to our three infantry battalions which were also a permanent part of the First Brigade, 4th Infantry Division:

- Alpha Battery direct support to 1st Battalion, 8th Infantry.
- Bravo Battery direct support to 3rd Battalion, 12th Infantry.
- Charlie Battery direct support to 3rd Battalion, 8th Infantry.

Additionally, the 29th Artillery Regiment in Vietnam included *Searchlight* batteries that employed 105mm howitzers with visible and infrared searchlights in support of infantry and other units in

---

[24] *The biscuits were delicious.*

several different areas of South Vietnam. Apparently, however, only one or two guns of one of the batteries were with the First Brigade area of operations in 1967 and '68.[25]

Other artillery units attached to the First Brigade provided what was called *general support*. They had larger caliber guns not permanently assigned to any specific infantry battalion but instead provided artillery fires as needed to all units, large or small, on demand. The general-support artillery units included the following:

- Alpha Battery, 5th Battalion, 16th Artillery (155mm howitzers).
- Delta Battery, 5th Battalion, 16th Artillery (8-inch howitzers).
- Bravo Battery, 6th Battalion, 14th Artillery (175mm guns).
- Charlie Battery, 6th Battalion, 14th Artillery (175mm guns).
- Charlie Battery, 1st Battalion, 92nd Artillery (155mm howitzers).

As the need arose in the everchanging environment, all batteries (direct-support as well as general-support) provided artillery support to all in need and as they were able.

During the first two weeks at Camp Enari we were abruptly awakened twice in the wee hours of the morning and ordered outside and into a large and partially underground sandbag bunker with a small movie projector and makeshift screen to keep us occupied until the all-clear was sounded about an hour later each time. I do not recall ever hearing any incoming mortar or rocket rounds while at Camp Enari, then or anytime thereafter, but the complex was large enough that something could easily happen in one location without being heard in another.

---

[25] *I have been able to find only limited information on the 29th Artillery Searchlight batteries operating in the Central Highlands of South Vietnam and hope to learn more.*

The 4th Infantry Division

No effort such as this would be complete without saying something about the long and distinguished history of the 4th Infantry Division of the United States Army which is frequently celebrated in military circles but much less known elsewhere due to a longstanding tradition of refusing to boast. Therefore, following is an abbreviated history for your reading enjoyment with parts borrowed directly and with permission from the archives of the 4th Infantry ("Ivy") Division Association of which I have been a member for many years.

On 10 December 1917 Major General George H. Cameron took command of the brand new 4th Infantry Division at Camp Green in North Carolina right after the United States had entered WWI. Roman numerals had long been used to designate military units, so the new division became "IV" and was quickly called the "Ivy" division. The new insignia was formed by four ivy leaves in the shape of a cross with the four green leaves overlaid on a khaki-colored background and the square patch displayed diagonally as a diamond. Our division motto became "Steadfast and Loyal," two words traditionally tied to the ivy plant.

4th Infantry Division Shoulder Patch on black background.

Just six months old, the 4th Infantry Division went into action in the Aisne-Marne campaign in July 1918 then fought the German Army in the American, British, and French sectors to earn five battle streamers. After WWI, the 4th Infantry Division was deactivated.

On 1 June 1940, the 4th Infantry Division was reactivated at Fort Benning in Georgia just eighteen months before the United States entered World War II when the Japanese attacked Pearl Harbor on 7 December 1941. After training for amphibious landings, the Ivy Division moved to England in January of 1944 to prepare for Operation Overlord, the D-Day landing at Normandy that began at 0630 hours[26] on Utah Beach, 6 June 1944 as the men of the 4th Infantry Division's 8th Infantry

---

[26] *Military time is expressed in hours rather than using AM or PM with the first twelve hours being identical to the numbered hours on the face of a clock and the last twelve hours expressed in numbers added to twelve. For example, 1:00pm would be 1300 hours (thirteen-hundred hours), 6:00pm would be 1800 hours (eighteen-hundred hours), and 10:00pm would be 2200 hours (twenty-two-hundred hours). Partial hours are expressed in hours and minutes so that 9:14pm*

Regiment were the very first ashore, accompanied by the assistant division commander, 56-year-old Brigadier General Theodore Roosevelt III who was not in peak health and walked with a cane. Against his better judgement, the division commander, Major General Raymond O. Barton, had finally granted Roosevelt permission to go ashore with the first wave of troops he had helped train, but things seldom go as planned. The English Channel was particularly rough that day with strong tides and currents, and the first assault boats were driven away from the intended landing area about a mile too far south. When the boats landed in the wrong place, Roosevelt took direct command and made the decision that the invasion was going to proceed from the point of landing and he immediately began organizing the men on the beach to move inland. Roosevelt is quoted as saying, "We'll start the war from right here" and to the horror of his staff, he hobbled unprotected upon the beach with a .45 automatic pistol in hand while shouting orders like the soldier he was. Despite being the son of the famous president of Rough Rider and San Juan Hill fame, Roosevelt had seen combat during WWI and knew exactly what he was doing. They had caught the Germans off-guard by landing at a location that was more lightly defended than the original landing point. Utah Beach was the first to be taken and with lower losses than the nearby units assaulting bloody Omaha beach on the same day. As Roosevelt pushed inland, the remainder of the 4th Infantry Division quickly followed and fought to Cherbourg, arriving on 25 June 1944, then they broke out through hedgerow country at St. Lo.[27]

The 4th Infantry Division was the first allied unit to enter Paris on 25 August 1944, then rapidly moved through northern France and reached Belgium and the German border the following September. Men of a very small 4th Infantry Division Recon unit were the first to cross onto German soil on 11 September 1944. Two months later in November, the division moved into the Hurtgen Forest where they fought some of the fiercest battles of WWII.

The 4th Division defended Luxembourg during the Battle of the Bulge and then fought back across Germany.[28] When the war in Europe ended on 8 May 1945, the 4th Infantry Division had participated in every campaign from Normandy, across France and deep into Germany. During

---

would be 2114 hours (twenty-one-fourteen hours or two-one-one-four hours). Midnight is expressed as 2400 hours (two-four-hundred hours) and 12:20am would be 0020 hours (zero-zero-twenty hours or zero-zero-two-zero hours).

[27] *For his actions on Utah Beach at Normandy, Theodore Roosevelt III was posthumously awarded the Medal of Honor after dying of a heart attack on 12 July 1944 and is buried there among his men.*

[28] *Lt. Gen. George S. Patton, Jr., wrote the following to Major General Raymond O. Barton, then the 4th Infantry Division Commander: "No American division in France has excelled the magnificent record of the 4th Inf. Div., which has been almost continuously in action since it fought its way ashore on the 6th day of last June; but in my opinion your most recent fight, when such a depleted and tired division you halted the left shoulder of the German thrust into the American lines and saved the City of Luxembourg, is your most outstanding accomplishment."*

WWII, five more battle streamers were added to the division colors as the division suffered more than 34,000 casualties.

The next call to arms was to Vietnam in August 1966 when the 4th Infantry Division began its four-and-a-half-year deployment into the Central Highlands. They were given several tasks including defeating the NVA in the Tri-Border area and blocking him along major infiltration points along the Ho Chi Minh Trail by way of Laos and Cambodia.[29]

Throughout most of my tour of duty, except all of March and part of April 1968,[30] the 4th Infantry Division was headquartered at Camp Enari with responsibility for the largest operational area of the war in what was called Two-Corps (II Corps) that induced much of the Central Highlands and under command of Major General William R. Peers. The division consisted of three brigades, each comprised of infantry, artillery and support units that were moved from one brigade to another as necessary, also sometimes being placed under temporary operational control (OPCON) to others as also occasionally happened at the brigade level as the various divisions and other larger organizations swapped men and material as needed.

By the time the 4th Infantry Division returned stateside to Fort Carson, Colorado at the end of 1970, more than 17,000 of our soldiers had been wounded or killed in combat and eleven 4th Infantry Division soldiers earned the Medal of Honor.

During the Vietnam War, eleven additional battle streamers were added to the 4th Division colors for combat action in places unknown to most Americans or now forgotten except by those who fought there.

In later years, the 4th Infantry Division would go on to serve in combat in Operation Iraqi Freedom, Operation Enduring Freedom, Operation New Dawn and, by the time this book is published, probably other operations, actions or "conflicts" as they are sometimes called by the people who don't have to fight them.

---

[29] *Parts of "the trail" reportedly ran from southern Cambodia northward into South Vietnam but most infiltration was from the north.*

[30] *During the brief time the 4th Infantry Division was officially headquartered at Dak To but with some of its support components still located at Camp Enari near Pleiku.*

Our Enemy – the NVA and VC

Those I spent time with were mostly of the enlisted ranks and junior NCOs and we called our enemy by many different names, some of which can be repeated in polite company and others that should not.

The North Vietnamese Army was officially called the People's Army of Vietnam (PAVN) but we called him "*NVA,*" "*Charlie.*" or sometimes *Regulars* (to distinguish him from the Viet Cong who were our other enemy.) While the NVA were the regular soldiers, the Viet Cong were the "irregulars" or guerrillas that were part of a noble-sounding organization called the National Front for the Liberation of South Vietnam, but we called them *VC* for short. Although I saw the NVA as soldiers, I thought the VC were nothing more than terrorists because of their indiscriminate methods which commonly included killing civilians – the people they were supposedly fighting to "liberate." In conversation and radio communication we called our enemy V*ictor-Charlie* (representing the letters *V* and *C* in the phonetic alphabet) which was shortened to *Charlie* and used to describe both the VC and NVA. The designation *Charlie* was used to describe either an individual enemy or any size enemy unit.

The NVA soldiers in our AO were for the most part well trained and motivated and we were told that some of them our age had been in combat since before we graduated high school which was probably to bolster our respect for them and lessen our youthful zeal. We were also told that, like us, their ranks included conscripts as well as volunteers and that many of them completed their training even while migrating from North Vietnam down through Laos and Cambodia into South Vietnam.

The NVA soldier was taught that his mission was to liberate the South Vietnamese from the government they had freely chosen. We, on the other hand, believed that the Communist North Vietnamese, with support from the Chinese and USSR, wanted to conquer the South against the will of the South Vietnamese people. The United States Congress beginning in the 1950s had supported the democratically elected governments of South Vietnam and had pledged to back them militarily if the need arose, which we did.

In the field, the North Vietnamese Army was comprised of uniformed soldiers who were usually very well equipped, many carrying the full-automatic Kalashnikov rifle (AK-47) or the lesser SKS, both of Soviet design and firing the same 7.62x39mm cartridge. As with American soldiers, the NVA foot soldiers were trained in many different specialties and employed a multitude of weapons including shoulder-fired and emplaced machineguns of several calibers, various types of hand grenades, B-40 and RPG rockets, flamethrowers, satchel charges, and other weapons. Others were trained in the use of 60mm and 82mm mortars, 122mm rockets, 57mm recoilless rifles, and 100mm artillery pieces. They also used other weapons, some captured from the French in the 1950s including .30 caliber M1 Carbines and other American made weapons. They were organized into typical military units including divisions, battalions and brigades.

The NVA and VC also knew how to make use of unexploded bombs and artillery shells as improvised explosive devices for mines and booby traps on roadways and elsewhere.

Some of the weapons, AK-47 magazines, 82mm mortars, concussion grenades, and other equipment that was carried by about twenty NVA who were all killed at Tan Canh in early February 1968 at the beginning of the Tet offensive. (Steve Stark photo)

Perimeter Guard Duty at Camp Enari

Every two or three days after arriving at Camp Enari and into late September 1967, we all took turns on twelve-hours nightshifts in four-man, above-ground permanent bunkers on the outer perimeter at Camp Enari. Bunkers were simply sandbag-encased wooden frames having poured concrete floors. They were about seven feet high on the inside, probably a dozen feet wide and roughly eight feet deep. I was always assigned to the same group of bunkers that covered an area about an eighth of a mile wide and a hundred meters deep with vegetation cut to under a foot high and having no trees. The area to our front, the kill zone, was a free-fire area where anything that moved at night was fair game. The rules of engagement were simple, and we were told that the local civilians had been repeatedly warned to stay away from the perimeter especially at night where mostly newly arrived (and undoubtedly uneasy and perhaps a little trigger-happy) GIs were on duty. Beyond the open area to our front were mostly low trees, light jungle, and tall bamboo thickets. Everything to our rear and on both flanks was inside the wire.
The bunkers were spaced in a row maybe fifty to sixty meters apart with command bunkers on stilts twenty-five, or so, feet off the ground located every three or four of the lower bunkers. Immediately in front of all the bunkers were several layers of permanently placed concertina wire where trip flares and claymore mines had been placed, their detonator wires running to the closest bunker. GIs before us had followed the longstanding practice in tying empty tin cans on the concertina to make noise when enemy sappers tried to sneak in or cut the wire.

It was possible to walk through the maze of concertina wire layers by following narrow and staggered openings but only in daylight when the paths were visible, even then often getting nicked by a sharp barb. After dark, everybody stayed inside the bunkers while taking turns in three-hour shifts so that two men were always awake. Although each bunker had a field phone, the officers and NCOs on duty still made inspections by Jeep a couple of times each night, arriving on the narrow single-lane dirt road to our rear that ran behind the bunkers. We could hear them coming (Jeeps are noisy) as they dismounted and approach on foot when we would sternly halt them to demand the password or countersign. They would sometimes stay around for a few minutes to check things out, ask a question or two and then depart to another bunker further on down the road. If an officer was performing the inspection, we had orders not to salute.[31]

In addition to our helmets and flak vests, each man carried an M16 and a pistol belt with magazine pouches, canteens and fanny pack attached. The bunker I occupied was supplied with fragmentation grenades packed in cardboard cylinders. Also, in the bunker was a supply of small rocket-propelled aerial parachute flares[32] that we called *knee flares* although we launched them by

---

[31] *The only time I remember saluting while in country was in our battalion area at Camp Enari and never at Dak To or any of the fire bases. Saluting was discouraged because it clearly identified the officer who then became a prime target.*
[32] *Officially the "Signal Illumination, White Star Parachute M127A1."*

striking them on a sandbag rather than our knee. We also had an M79 grenade launchers and two bandoleers of 40mm HE shells. Some of the bunkers included M60 machineguns but not ours.

At night as the bamboo cooled, it made a cracking or popping sound that scared hell out of me the first time I heard it because I thought the entire North Vietnamese army (undoubtedly, I imagined lead by Ho Chi Minh himself), was coming straight for my bunker. Nighttime in the Central Highlands was usually noisy thanks to the birds, monkeys, and goodness-knows-what lurking in the bush. One little animal was called the *fuck-you lizard* because it made a sound exactly like that. Another bird, monkey, bug or something seemed to call out *Charlie* all night long while other critters made noises that GIs interpreted in different ways, often vulgar. Early-on, we scoffed at rumors that guys on the perimeter sometimes reported hearing tigers roar,[33] and occasionally even elephants. Later we learned that both animals and others were all over our AO.[34]

One common rumor was that when the jungle got quiet, something bad was getting ready to happen. But another rumor said exactly the opposite – the birds and other animals made noise when people or badass creatures were moving around in the bush. In Vietnam the stories and rumors were never-ending, and it was often difficult to separate the wheat from the chaff. Even today, our infantrymen who fought right in the middle of it have widely different opinions about a lot of things.

When we heard (or thought we heard) noises at night on the perimeter we were permitted to fire knee flares to see what was out there if we didn't overdo it, which was a subjective call. Watching a flare float to the ground under the small white parachute offered a view of the open field for about twenty seconds which was reassuring. We never were ground attacked at Camp Enari while I was there, but the Pleiku area would witness ground attacks just four months later during the Tet Offensive of 1968.

---

[33] *On 10 January 1969, about three months after I came home, a First Brigade soldier with Alpha Company, 3rd Battalion, 8th Infantry was manning a night listening post at LZ Penny west of Kontum City and was attacked by a tiger and dragged away about 700 meters. They found his beheaded body the following day.*

[34] *AO is an abbreviation for Area of Operation.*

Easy Duty at the 6th Battalion, 29th Artillery Headquarters

One afternoon following all-night guard duty and after noon chow I was told to report to headquarters. There, I was interviewed by a Lieutenant who said he had read my 201-file[35] and noticed that I had worked for a cartographic drafting company in St. Louis on summer break throughout high school. He laid out a topographic map, pointed to a general area and asked me to explain what I would expect it to look like on the ground which I did. He asked a few more questions and I apparently passed his test because he immediately put me to work plotting locations on larger maps covered in clear plastic from several folded-up maps that were hand-drawn on notebook paper, most including roughly drawn trails and streams along with a few 4-place map coordinates and some roughly drawn contour lines, also with notations about tiny villages and vegetation such as bamboo and jungle density.

Over the next week or two between other assignments I worked the maps plotting coordinates and began understanding the geography of the area south and west of Camp Enari in Dar Lak Provence as well as in the Tri-Border Area north and west of us into Kontum Provence and out near the Cambodian and Laotian borders. It was rewarding work and I enjoyed being around everyone there from the highest-ranking officers on down to the Privates, like me. I overhead bits and pieces of conversations among the various officers and NCOs who came in for meetings wearing all kinds of uniform insignia for the different Army branches representing the combat branches – infantry, artillery and armor – as well as the support branches responsible for logistics, transportation, medical, air and having other responsibilities. It didn't take a military genius to figure out that something big was getting ready to happen.

At that time, some of the firing batteries of the 6th Battalion, 29th Artillery and much of the action was located at a place called Jackson Hole (sometimes called "Jackson's Hole") which was a fire base located about twenty-five miles generally west of Pleiku and near the town of Plei Djerang in Dar Lak Provence and there was great interest in that area and the country west of there and all the way out to the Cambodian border.

I soon learned that plans were being made to move us to Buon Ma Thuot[36] located roughly 125 miles south of Pleiku in Dar Lak Province where the NVA were expected to make their next big push into South Vietnam from Cambodia with support from others already in the neighborhood and concealed in the mountains and rain forests of the Tri-Border Area.

---

[35] *US Army 201 File is the equivalent of a personnel file.*

[36] *Spelled many other ways including sometimes Ban Me Thout.*

The plan, however, changed as fresh intelligence suggested that Dak To, further north, would be the next hot spot. As it turned out, the intelligence reports were correct.[37]

I turned twenty-one on 10 October 1967 at Camp Enari and my birthday celebration included getting ready for convoy duty the next day and then enjoying a warm can of Lone Star beer smuggled-in by one of my buddies. That evening I wrote my Dad to tell him I'd had my first legal drink.

---

[37] *Bob Quinn does an excellent job of detailing those events in his book, "The Battle of Dak-To Steadfast and Loyal' which is a must-read for anybody interested in Vietnam War history.*

Convoy Duty

Most enlisted men in our battalion and everybody in our barracks pulled convoy security duty every few days with no regular schedules and depending on other assignments. They announced names the day before the convoy was scheduled to leave Camp Enari.[38]

Convoys delivered ammunition and weapons of every type, and size and caliber including artillery munitions, grenades, ammo boxes full of 5.56mm bandoleers and 7.62 M60 belts, LAW rockets, Claymore mines, and everything in between. Also, they delivered the mail, food, truck parts, fuel, howitzers, artillery components, C-rations, sundry packs, corrugated steel sheeting, PSP, concertina wire, plywood, sandbags, tents, shovels, axes, sleeping bags, rope, ammunition slings, medical supplies, beer, and everything else needed in the field.[39]

Convoys were comprised of single-file lines of two and one-half ton ("deuce and-a-half") trucks, smaller three-quarter ton trucks, Jeeps, and often several types of armored vehicles. Occasionally we would see armored vehicles with twin-forty guns ("*Dusters*") and the awesome quad-fifties on the sides of roads looking for trouble.

While convoys slowly snaked through the valleys and jungle along dirt roads, UH-1 ("Huey") helicopter gunships sometimes provided protection by circling overhead in large oval patterns, occasionally two of them working together and running generally along with the convoy at elevations probably between 200 and 500 feet off the ground so they could keep an eye on the surrounding terrain and quickly respond as needed. We could easily see the faces of the helmeted door gunners constantly looking down and searching for trouble – always a comforting sight.

When we had a PRC-25 radio with us, we could easily distinguish the helicopter pilots from others because their voices would seem to vibrate as they spoke which was caused, I am told by an expert, by a phenomenon called RFI (radio frequency interference) caused by the rotor blade.[40]

---

[38] *The earlier convoys in October were between Camp Enari and fire bases located to our south and west, just east of the Ia Drang Valley. Later on, our convoys traveled primarily to the north and west to Dak To in Kontum Provence and out to the Ben Het area that was reachable by road during good weather.*

[39] *Helicopters delivered men and material where there we no roads or during the rainy season when they were impassable, or where bridges were damaged and not yet rebuilt. Transport planes were also used but most of the resupply was by truck convoy.*

[40] *In the movie "Good Morning Vietnam," Robin Williams does an excellent job of imitating the sound of a helicopter pilot by quickly punching his chest while talking about seeing water buffalo and "cooking them down" with napalm.*

When the Hueys flew low enough, their exhaust fumes mixed with the burning truck diesel and the dust and other jungle smells, often including the sweet fragrance of burning wood from a nearby Montagnard village. The Central Highlands was a place of many smells which is something no book or movie can ever capture but which every man who served there remembers.

While riding convoy duty during the dry season, the dust and sweat and diesel clouds mixed into a sort of paste-like goo that burned the eyes and impregnated skin pores. During the rainy season, everything and everyone was damp or soaked to the bone and the roads often became slick and almost impassable when stream crossings or mudslides melted them away.

In some places the vegetation had been cleared away from the roads by Army Engineers to maybe fifty meters, sometimes only twenty-five or less. At higher road elevations, and depending on the terrain, you could sometimes see for several hundred meters and maybe a mile or two out over the countryside. At other locations, the jungle was right on top of you making it impossible to see more than just a few feet past the road into the dark green and black shadows that hid everything inside even during the brightest midday sunshine. Moving slowly along, especially during the late afternoon, the jungle shadows danced around and played tricks on your eyes. Everybody knew we were sitting ducks, but nobody talked about it.

Although it was not my job on convoys, I usually carried a map and compass to track our progress because I wanted to know where we were just in case. Even so, I always believed that the officers and NCOs in charge knew exactly where we were along the way and had radio communication with those who would quickly come to our aid if attacked.

Convoys to and from Camp Enari varied in size and I never remember seeing one in its entirety due to the way they were formed. On the way out from Camp Enari, vehicles from various units near Pleiku would join up at some predetermined time and location, then be staged by whoever was in charge. As they began moving, the objective was to keep going at a steady pace of perhaps fifteen to twenty miles per hour without interruption if possible. Vehicle spacing was probably based on Army protocol or more probably as decided by the Convoy Commander and depending on the weather, intelligence reports, and his best judgement. A lot of the decisions made by our officers were judgement calls made on the fly which is what the big bucks were for.[41]

The trip between Camp Enari and The Oasis usually took a little over two hours while the later trips up to Dak To and vicinity took between about three and four hours. Always very long and

---

[41] *Second Lieutenants earned $321.00 per month and most Captains earned around $500.00 per month. I was a Private E-2 at that time earning $106.20 per month. Everyone, of course, also earned combat pay of approximately another $50.00 per month. All the above pay was before taxes.*

tiring, it was a relief to safety arrive at our destination despite the work yet to be done in helping offload whatever we were hauling.

On the later convoys to Dak To, I remembered occasionally stopping near Kontum City on the way up from Camp Enari. However, Tim Tyler, who drove convoys to and from Dak To much of his time in-country, says he remembers the trips from Camp Enari to Dak To were always nonstop but that occasionally, on the way back to Camp Enari, they would stop at Kontum City for the night if they had departed too late to make it all the way back by nightfall. The convoys from Dak To back to Camp Enari and the Pleiku area were usually smaller and often had lighter loads but would still be a tempting target.

When I was assigned to convoy duty, they put one man inside the truck with the driver and two men in the back with whatever we were hauling. As I recall, the vehicle cabs were typically covered but the payloads in back were not. I preferred riding in the open back for the freedom of movement as compared to being inside although it was not nearly as comfortable.
Roadside ambushes and snipers were always a concern, so most everybody stayed alert and wore their steel helmets. We hated wearing flak vests because they were hot and uncomfortable,[42] but usually wore them anyway on convoys. We knew that little could be done about land mines and improvised explosive devices and only hoped that the Engineers who regularly swept the roads had done their job well.

On convoy security duty hunkered down in the rear of a three-quarter-ton truck in open country on the way to The Oasis from Camp Enari on a sunny afternoon in mid-October of 1967. (Steve Stark photo)

---

[42] *Four months later in a letter home dated 11 February 1968, Major Leslie Crouse, our Battalion S3, grumbled to Mrs. Crouse that soldiers were not wearing their flak vests and that he believed that ninety percent of casualties did not have to happen.*

During a late afternoon convoy from The Oasis to Camp Enari in mid-October, our deuce-and-a-half loaded with 105mm ammo began to sputter and the driver pulled off the road as a smaller truck and the men aboard got out to help us.

The road was unusually straight, and we watched the remaining vehicles in the convoy pass us by and fade into the distance and out of sight. The terrain was flat with not much space between the road and the jungle, so the seven of us formed a perimeter around both trucks and lay down in the damp ditches on both sides of the road. We waited and watched as the shadows grew longer and the sun set – a somewhat lonely feeling. All was quiet until the birds and monkeys and other jungle critters decided to serenade us with their usual insults. Finally, maybe an hour later but what seemed like an eternity, an armored personnel carrier accompanying a deuce-and-a-half full of GIs and a much larger truck arrived to tow our truck the remainder of the way back to Camp Enari. Convoy security duty in the Central Highlands was thankless work; tiring, dirty, and sometimes dangerous. You will never see an award-winning movie or read a bestselling book about the guys who loaded and drove the trucks, humped and hauled the ammo and supplies through some of the most perilous places in all of Southeast Asia at the time. Never underestimate their contributions to the war.

The One-Jeep Convoy

Early on the morning of 26 October 1967 after riding convoy duty the day before, I was ordered to report to Sergeant Major Charles A. Taylor at battalion HQ who said he had been told that I was good at reading maps and knew that I held an Artillery Survey MOS. But he wanted to know if I also knew how to drive a Jeep and do field reconnaissance work. I told him, confidently, that although I had never driven a Jeep, I thought it would be about the same as driving a car or pickup truck and would not be very difficult. I quickly added that they had covered reconnaissance during AIT at Fort Sill in the classroom and on the West Range which was pretty simple – just look at the terrain, take compass readings, and compare them to the map. "No problem. I can handle it."[43]

Obviously not impressed with my answer, he then asked if I had a driver's license and I told him that I did, but it was in the connex. He asked, "Why is it in the connex," and I explained that we had been told to put our civilian identification and stateside clothing in the connex for safe keeping until it was time to go home and that's where my driver's license was located – my Missouri driver's license. Increasing the volume of his voice slightly, he told me in no uncertain terms that a Missouri driver's license was no good in Vietnam and that I needed an Army driver's license. He ordered me to report to Mr. Brauner[44] and get an Army driver's license adding, "...if you can pass the test."

Immediately heading for the motor pool area, I found Mr. Brauner working at his large metal desk with his assistant, a PFC, seated across from him. After dutifully repeating Sergeant Major Taylor's instructions, Mr. Brauner motioned toward a Jeep out front and told me to get in and drive it around the motor pool area, which I did. Almost immediately upon returning, his assistant handed me my Motor Vehicle Operator's Identification Card signed by Mr. Brauner which he told me to sign and keep with me while driving. Turning the card over, in addition to the ¼-ton M151A1 ¼ ton *truck* (commonly called a "Jeep"), I had also instantly been qualified to drive a three-quarter-ton truck and even a 2 ½-ton truck, the so-called *deuce-and-a-half* diesel rig – neither of which I had ever driven.

I wondered what had just happened and what my future held. Had the Army, for some unknown reason, suddenly and mysteriously decided in their infinite wisdom to convert me from an artillery surveyor into a convoy driver? In less than a month and for the second time since arriving in Vietnam I realized that my MOS meant very little and that the Army could (and would) use me as

---

[43] *Like most young men of my generation in their late teens and early twenties, I had all the answers.*

[44] *Norbert Brauner was our Warrant Officer (CW2) in charge of many things including maintenance and vehicles who was very highly respected because he really knew his stuff. Warrant Officers are not NCOs and not Officers but something in between and were addressed as "Mister" rather than "Sergeant" for NCOs or "Sir" for Officers. Many Warrant Officers were also trained as helicopter pilots and in other specialty areas.*

they pleased – they owned me. I had already been on several convoys and knew that the drivers were basically just slow-moving targets and that their job was much more dangerous than mine just sitting at a desk at base camp and plotting map locations while chatting and drinking coffee. The eleven months remaining on my tour of duty was beginning to seem further and further away.

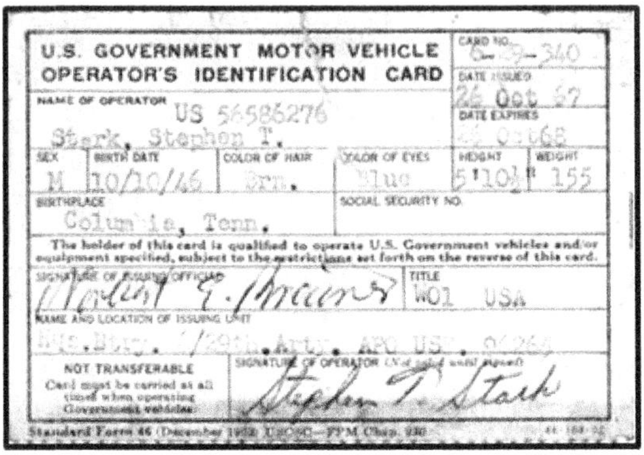
U.S. GOVERNMENT MOTOR VEHICLE OPERATOR'S IDENTIFICATION CARD

CARD NO. [illegible]-340

DATE ISSUED 26 Oct 67

NAME OF OPERATOR US 56586276 Stark, Stephen T.

DATE EXPIRES [illegible]

SEX M | BIRTH DATE 10/10/46 | COLOR OF HAIR Brn. | COLOR OF EYES Blue | HEIGHT 5'10½" | WEIGHT 155

BIRTHPLACE Columbia, Tenn.

SOCIAL SECURITY NO.

The holder of this card is qualified to operate U.S. Government vehicles and/or equipment specified, subject to the restrictions set forth on the reverse of this card.

SIGNATURE OF ISSUING OFFICIAL

TITLE WO1 USA

NAME AND LOCATION OF ISSUING UNIT

NOT TRANSFERABLE
Card must be carried at all times when operating Government vehicles.

SIGNATURE OF OPERATOR (Not valid until signed)

Standard Form 46 (December 1952) [illegible]

U. S. Government Motor Vehicle Operator's Identification Card

Returning to battalion headquarters, I reported to Sergeant Major Taylor and showed him my license. He quickly inspected it and then told me to go and "get a Jeep" and report back to him because we were going on a little road trip.

"Get a Jeep?" I thought. "How the hell does that work, especially for a Private?"

But once again I went jogging back down to the motor pool where they kept the Jeeps and again found Mr. Brauner and repeated Sergeant Major Taylor's instructions.

Mr. Brauner's able assistant did some paperwork, handed it to me and said, "Sign it" which I did. He motioned outside to the same Jeep used for my driving test and I got in and drove back up the hill to headquarters and found Sergeant Major Taylor. He stepped outside, pointed to the Jeep and complained that it was dirty and said I needed to "clean it up." Then, he said, I was to collect my rifle, flak vest, and all my field gear and report back to him. Returning to the motor pool for a third time (but this time driving), I remember thinking that I had not seen a Jeep or for that matter anything else during my short time in Vietnam that would even come close to meeting the definition of "clean."

Mr. Brauner's assistant, (by now probably getting a kick out of the whole episode) pointed to a bucket outside and a water hose on the ground in the mud. I found an old Army brush and some almost clean rags for the windshield and brushed the wheels and tires, also spiffing up the interior as best I could. After checking the oil, fan belt, tires and spare, I decided everything was good to go and drove the few yards to our barracks just across an open area from the motor pool to gather my field gear, donning my flak vest and helmet placing everything else in the back seat of the Jeep, including my rifle.

Standing outside HQ were Sergeant Major Taylor and Master Sergeant Billy Stout, both with M16s shoulder-slung and a pile of gear on the ground next to them including field packs, pistol belts, gas masks, and canteens. Sergeant Major Taylor and Sergeant Stout, who were both of considerable bulk but not fat, had on flak vests that were completely zipped up and snapped; also, with OD towels wrapped around their necks with steel pots pulled down tight which made them both look like two great big turtles. I'm sure I grinned when I saw them but, of course, said nothing.

In addition to their rifles, both carried 1911 forty-fives in Army flap holsters on pistol belts with magazine pouches attached. Sergeant Stout stowed most of their gear in the back-seat area and climbed in on the right side as Sergeant Major Taylor got in the front seat and placed a second pistol belt and his M16 bandoleer on the floor.

Sergeant Stout handed Sergeant Major Taylor my M16 from the back where I had put it and Sergeant Major Taylor handed it to me. The Jeep had no rifle scabbard or bracket so I tried tucking it next to my right leg but that obviously wasn't going to work. Sergeant Major Taylor seemed entertained at my effort but calmly, almost kindly, asked me for my rifle which he placed on the floor next to his gear and sort of secured it with his left foot. He told me that if I needed it, he would get it to me right away and I believed him. I am certain that he realized I was a little apprehensive.

Sergeant Major Taylor motioned toward the road that ran past our battalion area and we headed out. At the main gate at Camp Enari, he told the MP in charge that we were headed out to join a convoy just up the road and that we were already running late. The MP asked no questions and immediately waved us through, and I drove past another smaller checkpoint to link up with the convoy. Once out of camp, Sergeant Major Taylor put a magazine into his M16 and I heard Sergeant Stout do the same thing, and they both then chambered a round.[45]
About a half-hour later there was still no sign of a convoy and it suddenly dawned on me that maybe there was no convoy – perhaps there never was. Had Sergeant Major Taylor lied to the MPs at Camp Enari? Was I the only one in the Jeep who had no idea what was going on? But I said nothing and continued driving north over the rough dirt road where we occasionally passed a building or small hamlet, also seeing a few Vietnamese civilians on foot or in motorized carts.

During the entire trip, Sergeant Major Taylor was the perfect backseat driver who continually complained about my driving and regularly reminded me not to run over the water buffalo dung

---

[45] *Like most fire bases, the standing order at Camp Enari was for rifles to be carried unloaded and without magazines. I later learned that even some infantry units in the field did not allow chambering a round even while on patrol except under certain circumstances.*

and to miss the chugholes. He said more than once that Charlie was very skillful at camouflaging land mines and booby traps and that the NVA had snipers that were much better shots than the ones he knew in Korea back in the '50s he said, "couldn't hit shit." The message was getting through and by then my pucker factor was probably approaching eighty-five or maybe even ninety percent.

Sergeant Stout probably heard parts of Sergeant Major Taylor's lectures over the noise of the Jeep, but he added nothing although he and Sergeant Major Taylor occasionally communicated among themselves as I concentrated entirely on my driving while carefully shifting gears and doing my best to miss what I imagined as mines while speeding up on the straighter and smoother stretches of road.

We arrived near Kontum City and stopped briefly for a water break near a roadside bunker where several sloppy-looking ARVN soldiers were lounging around, just talking and smoking cigarettes. Continuing north, we pulled off the road to allow a small convoy headed south to pass us.

Up the road further, we arrived at the tiny village of Tan Canh[46] where Sergeant Major Taylor motioned me to turn left and we headed west, first passing a small bullet-pocked concrete building by an airstrip[47] on the right, then onward about a mile further and stopping on the south side of a second larger airstrip. We had arrived at the Dak To forward base. I was not impressed.

Sergeant Major Taylor and Sergeant Stout climbed out and stretched and so did I. Sergeant Major Taylor said they had business to attend to and for me to get my gear out of the Jeep and find a way back to Camp Enari, but he did not say how. I separated my equipment from the pile in the back of the Jeep and collected my rifle. Sergeant Stout took the driver's seat while Sergeant Major Taylor climbed in and they drove away in my Jeep, and toward a cluster of sandbag-encircled tents and a large bunker to the south and a little west of the airstrip.

A Huey and OH-23 were parked on the south side of the airstrip, both inside separate short-walled sandbag revetments. Nearby were a few small tents also individually encircled by stacked sandbags. Just to our east were two or three above-ground bunkers like the ones on the perimeter at Camp Enari but smaller. Over the flat landscape in all directions except to the north across the airstrip were other bunkers and tents scattered around the area mainly to the south and back east toward Tan Canh. To west were a few small lean-to foxholes but not as many as elsewhere. Although I do not remember any bunkers on the north side of the airstrip, there must have been

---

[46] *As with many place names in Vietnam, Tan Canh is found spelled in different ways including sometimes Tan Cahn.*

[47] *The smaller airstrip just east of Tan Cahn was only about 800 meters long whereas the larger one at what would become the center of the Dak To forward base was about twice that length.*

some especially considering the terrain and nearby hills. To the south and southwest on the other side of the river were mountains much higher than those in other directions.

Clearly, we were down in the bottom of a bowl – a mountain valley where it seemed obvious to me that nobody in his right mind would ever put a camp, or post or any other military installation where the enemy could easily rain all kinds of high-angle and even lower-angle artillery down on us. At that time the area around the Dak To airstrip was much less organized and more lightly defended than it would become just a few weeks later and into the following years as the North Vietnamese repeatedly attacked.[48]

Late that afternoon and into the evening I filled sandbags with the infantrymen already there to improve the overhead cover of two of the bunkers. Around dark, we ate C-rations heated on an improvised tin can heater then I found a cozy spot in one of the tents to stow my gear, make a few notes in my notebook and spend the night. I slept very well my first night at Dak To which passed without incident.

The following day, 27 October, I arose early and had another C-ration meal. Somebody had prepared boiled coffee in an aluminum pot, but it tasted so terrible that I instead washed down my breakfast with a warm can of Coke from my pack.

A Slick that landed around mid-morning to drop off supplies gave me a lift back to Camp Enari and I caught a ride back to the 6th Battalion, 29th Artillery headquarters to rejoin the Recon/Survey Section, never again to drive a Jeep or anything else during my entire time in Vietnam.
Master Sergeant Stout and Sergeant Major Taylor remained at Dak To for an additional day but then flew back to Camp Enari the afternoon of 28 October.

The following day, 29 October 1967, the 6th Battalion, 29th Artillery moved its headquarters to Dak To in a large convoy that also included part of the headquarters of the First Brigade, 4th Infantry Division. We departed Camp Enari at ten-hundred-hours and arrived just six and a half hours later at 1630 hours, which was only about two hours longer than our solo Jeep ride just three days earlier. As with most of Headquarters Battery, the Recon/Survey Section made the move to Dak To doubling as security and riding in the rear of several different trucks. That trip was much more pleasant than my first trip to Dak To and passed without incident.

Dak To would then become my new home over most of the next ten months during the remainder of my time in Vietnam.

---

[48] *Even eighteen months later the May-July 1969 timeframe the area was still under siege when soldiers of the 299th Combat Engineers defended the Dak To air strip and forward base, or what was left of it.*

Over the years I have often wondered why Sergeant Major Taylor and Sergeant Stout didn't simply take a helicopter to Dak To instead of riding in a Jeep. Flying, after all, would be faster and obviously much less hazardous and besides, I could have stayed back at Camp Enari at my comfortable table working on maps rather than worrying about snipers and landmines.
Perhaps Sergeant Stout and Sergeant Major Taylor, who were the two most experienced and senior NCOs of our battalion, just wanted to see the terrain between Camp Enari and Dak To first-hand and from the ground rather than from the air. Both men, Stout now age 37 and Taylor 38, had seen combat during the Korean War and clearly understood the undisputable value of good ground reconnaissance.[49]

One Saturday morning almost forty years later I received a telephone call at home and the gruff voice on the other end said, "Steve, this is Andy Taylor" and all I could think of was Sheriff Andy Taylor from the Andy Griffith show – the one about Mayberry. Soon it was clear that the call was from Sergeant Major Charles A. (now "Andy") Taylor and it seemed very strange that he would call me, and especially by my first name, because we had been anything but *buddies* in Vietnam and the only personal contact I ever had with him was during that Jeep ride to Dak To and a very few other times when I was unable to avoid him. He told me that Ray Harton had given him my home telephone number and that he just wanted to talk to someone who had been with him at Dak To.

Sergeant Major Taylor once told me that he remembered that first Jeep trip from Camp Enari to Dak To but did not remember his driver, only that his habit was to "volunteer" someone for that kind of job. He said the trip was to select a suitable location for the 6th Battalion, 29th Artillery headquarters TOC and the chosen location was near the highest point of the camp and with a clear view to the south and southwest. He also told me that Colonel Donald Kenney, our battalion CO at the time, had told them he wanted to be located near the First Brigade headquarters TOC.

Over the following years until just before his death, either he called me or I called him every couple of months and we talked about all sorts of things including the war and what he always called the "Big Bang" on 15 November 1967 when the Dak To ammo dump was destroyed and earlier that same day when the Recon/Survey Section was performing a crater analysis and he and Sid Hegener had rushed to the fire to put it out when it first started.

---

[49] *It should also be noted that on 31 August 1968, Sergeant Major Charles "Andy" Taylor was with Major Leslie Crouse when Major Crouse was killed in action near Ben Het as they were preparing to perform a crater analysis to determine the source of ongoing numerous mortar and rocket attacks. Years later, Sergeant Major Taylor contacted the Crouse family, writing a letter to Mrs. Crouse detailing Major Crouse's death which she and her family appreciated greatly.*

Over the phone, I also spoke occasionally with his wife, Rose, the love of his life. Calling him one Christmas day, I also visited with his son, Jim, who was a career soldier like his dad and home on leave for the holidays. The Taylor family is full of patriots.

Miss Rose or "Biddy" as he often lovingly called her, passed away before Sergeant Major Taylor which was not in his plans and broke his heart. The tough old soldier had finally softened a little but would occasionally still revert to his gruff old self when he had a point to make. He frequently expressed his concerns over the future of the Army and the fate of our nation – he saw no clear distinction between the two. He worried also about the dwindling condition of our youth and the lack of manliness in boys who should have become men by now. He also occasionally used Bible verse to make a point which seemed strange since I had known him in his "rougher" days in Vietnam during a time when many of us did not like being around him.[50]

[50] *During a phone conversation he once asked my opinion of him while we were in Vietnam and I told him the truth which I think he already knew because it did not seem to surprise or bother him.*

The Central Highlands Tri-Border Area

In the Central Highlands approximately eighteen miles west of Dak To at the confluence of Laos, Cambodia, and South Vietnam was the Tri-Border Area – some of the most rugged terrain in all of Vietnam. It offered the perfect environment for the North Vietnamese Army to enter South Vietnam to hide, dig in, plan and then launch rocket, mortar, recoilless rifle, artillery and ground attacks against U. S. Army units and our allies in the area and even beyond.

The terrain consisted of forested mountains, many reaching elevations of more than 4,000 feet and covered with large trees and jungle with valleys and flatter grassy areas intermittently covered with trees more than one-hundred feet in height and dense bamboo thickets regularly rising forty to fifty feet straight up. The higher elevations dropped quickly into valleys also of dense jungle with multiple small rivers and streams in a rain forest of perpetual green twilight where everything was forever wet or damp. There, mosquitoes and leeches thrived in abundance with tigers, venomous snakes and other enemies of man along with monkeys and every imaginable color and shape of bug and lizard. In the lowlands, daytime temperatures often sweltered into the mid-nineties while nighttime temperatures in the mountains dropped to fifty or below and often felt much colder due to the high humidity.

The first time the U. S. Army met a large force of NVA was in November 1965 in the Ia Drang Valley located roughly fifty miles south of Dak To where part of the recently-formed First Cavalry Division (Airmobile)[51], under the leadership of Colonel Harold G. (Hal) Moore, successfully stopped the NVA push to the east but at high cost. That action is chronicled in the book titled "*We Were Soldiers Once, and Young*"[52] written by Hal Moore and co-authored by an unusually honorable journalist named Joe Galloway.

NVA and VC forces returned to the Ia Drang Valley during the spring of 1967 resulting in heavy fighting near the Cambodian border in what is sometimes called the beginning of the "Border Battles," sometimes also called "The Battle of the Slopes" due to the terrain on which it was fought. One particularly long and bloody engagement at that time was in an area called The Valley of Tears where the First Brigade of the 4th Infantry Division first engaged the NVA during what is now known as the "Nine Days in May." There, the 3rd Battalion, 12th Infantry, with the 6th Battalion, 29th Artillery providing direct support, were the first units of the 4th Infantry Division to

---

[51] On 16 June 1965, the United States Air Cavalry was formed under authorization of the Defense Department as the First Cavalry Division (Airmobile) to improve combat mobility. The decision to form such a unit was a result of lessons learned during the Koran War and was not only the result of America's increasing involvement in the Vietnam War. Even so, timing could not have been better.

[52] The movie "We Were Soldiers" is entertaining but misses many of the important historical details included in the book.

engage the NVA. This is also where our very own Ray Harton earned his Air Medal with "V" as an artillery Captain while coordinating fire from a Huey gunship during heavy enemy contact.

As the fighting unfolded, Captain Harton learned about a stranded LRRP team in the vicinity that was in jeopardy of being overrun. After analyzing the situation, Harton rescued the LRRP team under fire and then returned to base to drop them off. The pilot, Harton and his crewmembers inspected the bullet-damaged Slick and were satisfied it would still fly, so Harton and his team quickly gathered water cans and ammunition, taking off again to fly back into the action, this time directly above the fighting to lower the ammunition and water by rope to the troops in the jungle below.

During that same general timeframe, the four First Brigade, 4th Infantry Division soldiers listed below earned the Medal of Honor. The letter P indicates they earned it posthumously.

- McNerney, David H., First Sgt., Alpha Company, 1st Bn., 8th Infantry (earned 3/22/67)
- Grandstaff, Bruce A., SFC, Bravo Company, 1st Bn., 8th Infantry (earned 5/18/67) P
- Molnar, Frankie Z., Staff Sgt., Bravo Company, 1st Bn., 8th Infantry (earned 5/20/67) P
- Bellrichard, Leslie Allen, PFC, Charlie Company, 1st Bn., 8th Infantry (earned 5/20/67) P

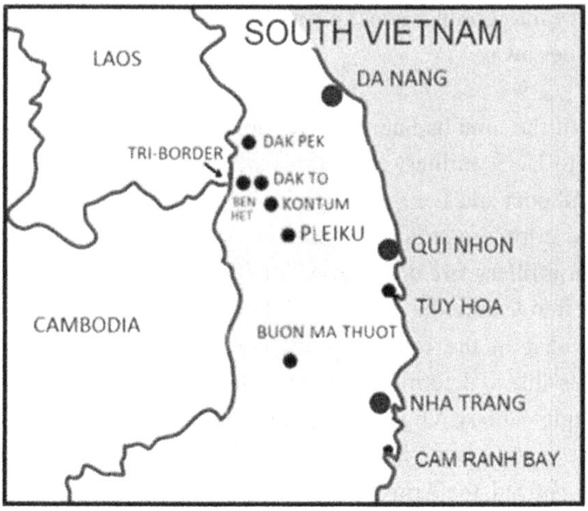

The Central Highlands (shaded) also showing confluence of Laos, Cambodia, and South Vietnam forming the approximate center of what was called the "Tri- Border Area."

In the summer months of 1967, the famous 173 Airborne had tangled with the NVA southwest of Dak To but with less than satisfactory results.

Now, in the fall of 1967, and almost exactly two years after Colonel Moore's now famous fight in the Ia Drang valley, some of the same NVA men and units were back once again in the Tri-Border Area, but this time in much larger numbers.

Dak To is sometimes considered a general area rather than a specific town or village or airstrip. Some maps of the area show villages labeled "Old Dak To" and "New Dak To" and nearby airstrips labeled "Dak To 1," and "Dak To 2." The entire compound, the *Dak To Complex*, was positioned in one of the larger valley areas on the north bank of the Dak Poko River. The center of camp was a little more than two miles west of the tiny village of Tan Canh at the confluence of Route 19 and Route 512 with a total area of about two-and-a-half square kilometers (a little less than one square mile). The First Brigade TOC was located near the southwest end of the 4,900-foot-long airstrip where observation helicopters, Huey Gun Ships, Slicks, Cobras, Bird Dogs, Air Force FACs, and other fixed-wing aircraft of every description including C-123 and C-130 transports regularly landed and departed.

The Dak To area was strategically significant for several reasons including its location. The NVA and VC depended on a broad network of roads, paths, and bridges called the Ho Chi Minh Trail that was constructed and maintained through the jungles of Laos and Cambodia and used to transport men, arms, and munitions for use against U. S. and friendly forces in the south. One of the main points of entry into South Vietnam from Cambodia was immediately to the west of Dak To less than fifteen miles away.

American politicians of the time had agreed with politicians of other nations that Cambodia and Laos were off-limits to U. S. military operations which meant the NVA could mass along the border just inside Cambodia and Laos and then move east into the Tri-Border area very quickly. This greatly hampered efforts to stop the continuing inflow of troops and materiel and we were forbidden to direct our artillery fire on the enemy until he crossed into South Vietnam. Even so, artillery fire missions into Cambodia and Laos did occasionally occur and often with the usually accepted explanation that in the rough mountainous terrain of the extreme western Central Highlands it was impossible to determine the exact location of the international borders especially in deep valleys and triple-canopy jungle where the maps were especially imprecise.

To help keep things organized, the Army gave a name or number (often both) to each of its manned locations and other places of interest. Fire bases were often identified by map spot elevations, in meters[53] and many fire bases and other locations had multiple names or nicknames depending on when they were occupied or by the occupying unit. The Dak To forward base was located near the westernmost of two airstrips and was known at various times by multiple names including

[53] *One kilometer in distance (often called a "click") is 1,000 meters or roughly 0.62 miles. One meter is about 1.1 yards, so that 100 yards (the length of a football field) is roughly 90 meters.*

Delta-Tango (D-T in the phonetic alphabet[54]), Fire Base 1, Dak To 2, Tan Canh Airstrip, and the Phoenix Airstrip. One of the reports later written by Lieutenant Colonel Albert Akers in February 1968 even calls the base "New Dak To." The eastern most and much smaller airstrip (which was outside the Dak To forward base perimeter) was sometimes referred to as Dak To 1, Tan Canh Airstrip, or simply Tan Canh (sometimes Tan Cahn, Tan Can or even Tin Can).

Often, a new fire support base (FSB)[55] was established when moving to a new location but sometimes old locations were reused, usually located on hilltops for providing fires to infantry units or overlooking anticipated routes of enemy travel. Although infantry and artillery units manned fire bases simultaneously, infantry units sometimes named them to their liking which occasionally caused temporary confusion as artillery and infantry liaison teams and others sorted things out. There was a lot of informal naming of almost everything and everyone in Vietnam.

Wherever the First Brigade infantry battalions went, the 6th Battalion, 29th Artillery followed while often occupying the same LZs and fire support bases. The infantry would first be lifted in by Slicks supported by Command and Control helicopters and gunships to clear the area and establish a perimeter. Then, sometimes at the same time but depending on the circumstances, an artillery advance party would join them to determine the layout of the fire base including gun placement, FDC, and to begin preparing for the first artillery pieces to arrive. Immediately thereafter, our 105mm howitzers, one at a time and slung underneath Chinooks (with a basic supply of 105 ammunition hanging below), were lifted in, often with the gun crews also aboard to immediately place and prepare the howitzers should a sudden need arise. Shortly thereafter, if requested by the Battery Commander, members of the Recon/Survey Section would arrive to survey in the battery. The exact sequence was determined by the Battery Commander.

By late October we were getting settled in as work continued to build and then improve the 6th Battalion, 29th Artillery TOC and other structures. The TOC had begun life as a sandbag-encased three-quarter ton truck with two radios aboard that was parked precisely on the spot selected by Sergeant Major Taylor and Sergeant Stout.

---

[54] *The phonetic alphabet consists of words used to depict letters to help avoid misunderstandings. The letters "b" and "d" for example sound similar but their corresponding phonetic alphabet letters "bravo" and "delta" are easily distinguishable ever when radio or other communication may be garbled.*

[55] *FSB is an abbreviation for "fire support base" but the word "support" was often dropped and the term "fire base" was more commonly used especially in conversations.*

6th Battalion, 29th Artillery headquarters TOC at Dak To in November 1967 looking southeast. Note the Recon/Survey parapet on the top right, BC scope protruding. (Steve Stark photo)

The battalion Commo Section guys laid landlines and set antennas for the radios and set two small gasoline-powered generators and an array of wet-cell batteries for the generators to keep charged, all encased in sandbag bulwarks for protection. We also constructed a sandbag parapet – a "watch tower" – on the west side of the TOC that was immediately manned by the Recon/Survey Section twenty-four hours a day to keep an eye on the surrounding hills, both close-in and far away. This is where we mounted our BC scope.

Dak To airstrip control tower November 1967 looking east. (Steve Stark photo)

At the same time, the very skilled 4th Infantry Division Engineers, with help from many "volunteers," worked to reinforce and improve the First Brigade TOC and the helicopter landing area next to the airstrip with additional PSP, also reinforcing the control tower with sandbags.

At that time, most everyone at Dak To was billeted in tents,[56] each one with sandbags stacked all around about four feet high and with a small opening at the front; some had blast walls. Each tent could hold about fourteen men and all their gear. Our officers had small round tents that slept three or four also surrounded by sandbags. Our tents had wooden cots, mosquito nets, and sleeping bags and we originally stored our survey gear inside but soon build a sort of lean-to right outside the entrance to provide more room.

When time allowed, we built wooden-plank floors inside our tent using disassembled artillery ammo boxes and sprayed them with diesel using Doc Rivera's hand-pump sprayer to help keep down the dust while also discouraging some of the local crawling critters from bothering us. We also built walkways around the battalion area from the same wooden boxes for dust control during the dry season and to hopefully make moving around camp easier during the upcoming monsoon season they said typically began in April or May each year.

In our spare time, we wrote letters home or read the latest hotrod or girlie magazines or participated in light gambling and several guys had pets including Sid Hegener had a parrot that would perch on his shoulder and eat from his hand. We also kept a pet rat in a chicken-wire cage for a while and fed it leftover mess hall chow. One of the guys in the nearby Commo Section had a spider monkey that ate Bit-O-Honey candy bars (paper and all) and drank warm beer when we had lots of it. We found our dog, naming her Sandbags wandering the camp and quickly took her in as our mascot.

Steve Stark shares a meal with Pro-Jo (left) and Sandbags. Everyone in our unit thought those two dogs were their personal pets – and they were. (Steve Stark photo taken by Gary Herder)

---

[56] *It wasn't until around mid-December 1967 that we began building sleeping quarters in underground bunkers.*

6th Battalion, 29th Artillery area at Dak To in early November 1967 looking due south down the hill toward the Dak Poko River and showing the partially completed wooden walkway. (Steve Stark photo)

We were living in high comfort after constructing a makeshift shower and a sort of steam-bath in the battalion area, both built under direction of our Lieutenant Skoglund. Given enough wooden ammo crates and nails we could build just about anything. That kind of work strengthened camaraderie and helped keep us fit and loose and our minds occupied.

During this time, additional bunkers were going up on the perimeters and the airstrip was continually busy as troop-packed trucks and armored personnel carriers moved through camp to points further west out to the vicinity of the Ben Het Special Forces camp which had grown into a sort of complex of four or five hilltops in the area. Twin-Forty guns on armored vehicles and Dusters also regularly came through Dak To on their way west with some remained with us for a few days remaining well inside the wire and obviously to counter anticipated ground attacks.

Sid ("Skid-Row") Hegener and his pet parrot at Dak To early November 1968. Sid would go on to earn an Army Commendation Medal with V during the Battle of Dak To. (Steve Stark photo)

The Recon/Survey Section at Dak To

We called ourselves simply "Surveyors" or the "Survey Section" and it was not until recent years that I learned from Bob Barrett and others that our leaders (our bosses, mostly Captains and above at the battalion and brigade level) referred to us as the Reconnaissance & Survey Section or simply "Recon/Survey." As it turned out, that name was appropriate because we did a lot more than artillery surveying – the sciences required in applying theodolites and other precision instruments to determine the exact location of artillery pieces on fire bases to give Battery Commanders a reliable beginning point from which his artillery could reliably deliver fire upon the enemy.

Most of the time at Dak To the Recon/Survey Section was without an officer in direct command, so we received orders and guidance from one of several NCOs including Staff Sergeant Clarence A. Price who always stood up for "his boys" and got us out of hot water more than once and sometimes when we probably did not deserve it. Glen Farasy once stored a small quantity of an adult beverage in our bunker and a junior officer from outside our Section somehow spotted it. When the officer inquired about it, Farasy told the truth – he was just storing it for someone else but could not (would not) divulge its owner. Sergeant Price was there during the "interrogation" and came to Farasy's rescue to keep the situation from getting out of control. At that time, relatively simple mistakes and minor infractions by recently hatched soldiers that would have been a big deal stateside or even at Camp Enari were often forgiven and never mentioned again thanks to the well-known camaraderie shared by U. S. soldiers in the field throughout history. There were times, of course, when things were not overlooked that resulted in demotions, punishment, or reassignments as deemed appropriate by our leaders.

All the guys in the Recon/Survey Section stuck together through thick and thin and never do I recall a fight or even a serious argument our entire time together. But like any other *Band of Brothers* we sometimes had disagreements that we always settled among ourselves.

From the fall of 1967 through 1968 our Recon/Survey Section consisted of somewhere between ten and about fifteen enlisted men depending on the rotation cycle and as guys arrived and departed.

In the beginning when we arrived at Dak To, our Recon/Survey Section consisted of at least the following ten men, all who had earned the Artillery Survey MOS at Fort Sill: Mike Alesi, Dave Drobac, Glen Farasy, Sid Hegener, Gary Herder, Ron Holloway, Pat Jones, Al Roden, Steve Stark, and Sam Weeks. Later, Sam was shipped out to the Americal Division at Chu Lai and never heard from again despite repeated attempts to contact him. There may have been others with us even during those earlier times and apologies are in order if I've forgotten anyone.

Many had nicknames that evolved in various ways.

- Mike Alesi was everybody's best friend and of proud Italian heritage and his nickname "WOP" in no way offended him or anybody else.
- Dave Drobac (later changed his name to Dave Foster) was "Rabbit," but nobody now remembers how that nickname came about.
- Ted Dieumegard, our Lieutenant beginning in February 1968 (after Larry Skoglund was wounded the second time),[57] was christened "Uncle Ted" but he never knew it until our reunion in Nashville in 2016. It probably had something to do with his very notable handlebar moustache.
- Glen Farasy got his nickname "Frog" from Al Roden who spotted Farasy peeking out from under a poncho liner early one morning wearing his glasses and Roden said he looked like a frog, so the name stuck.
- Sid Hegener's first name evolved into "Skid" and then "Skid Row" – we called him both.
- Gary Herder was the "old man" of our Section at age 25, so the nickname "Pops" fit him perfectly.
- Rob Homyak joined us in May 1968 having received his nickname "Slip" while stationed in Hawaii before arriving in Vietnam thanks to his heavy New York accent that sounded just like the character Slip Mahoney in the *Bowery Boys* (and it still does).
- Pat Jones was by far the largest and strongest among us, so he was "Moose."
- Harold Massey who also joined us at Dak To in '68 never had a nickname that I recall.
- Pat Reilly brought his nickname "Leper" from New York with him when he joined us in May 1968 – it's a very long story which Pat delights in telling at every reunion.
- Al Roden's nickname "Rat" was a spin on his last name and had nothing to do with his outstanding loyalty or character. Al had a way of keeping us laughing and lighthearted even during the darkest of times.
- Tim Tyler was "T-Man" and spent much of his tour driving convoy duty between Dak To and Camp Enari.

At one time or another during my year in country others who also served with the Recon/Survey Section included Mike Ferrell, Larry R. Flanagin, Ken Krieger, Larry Lightner (nickname "Flash"), Richard Mason, and Larry Maxwell.

Two times during my tour of duty there were what the Army called "infusions" where units exchanged men to mix the less experienced with those having more time in country. Also, this

[57] *Coincidentally, on this very same day and just before Skoglund was wounded (and very possibly on the same plane), Lieutenant Ted ("Uncle Ted") Dieumegard had arrived at Dak To. But, as with Lieutenant Skoglund, Dieumegard was quickly snatched up as a FO and spent several months in the field with the infantry before returning to Dak To as an Aerial Observation Officer in the fall of 1968.*

practice helped offset the cycling of troops arriving and departing at odd intervals. At various times we gained Rob Homyak, Roger Kvasager, Harold Massey, Pat Reilly, and Tim Tyler. In all cases, the new guys fit right in with the already tight-knit group. Ron Holloway was caught in an infusion in May of 1968 and ended up in the First Cavalry up north near the DMZ.

Our duties varied greatly and included the usual routine chores that all soldiers endure in addition to our "official" job of surveying-in firing batteries when required and performing other jobs fitting our training and skills. One of the nicknames given the Recon/Survey Section was "Crater Chasers" because one of our more interesting jobs was performing what were called crater analyses which consisted of locating recent mortar, rocket, or artillery craters and digging up the fuze or tail fin to determine the approximate direction from which the shells originated.[58] Time being of the essence, this chore was best performed as soon as possible after rounds landed which could be tricky because they sometimes impacted at odd intervals, often with several minutes between each volley.

---

[58] *When a mortar round impacts, the only thing remaining is the tail fin section that becomes buried in the earth at about the same angle as when it landed. Artillery rounds, similarly, leave the fuze as a telltale sign of the direction of its origin. It's a somewhat imprecise science and requires practice but can be quite effective in determining the direction of origin.*

Montagnards

Another of our duties, this one much more pleasant, was to accompany our battalion surgeon, Cpt. Raphael "Doc" Rivera-Rodriguez and his small team of Medics to nearby villages for a MEDCAP[59] to provide security while Doc and his crew tended the native people of the Central Highlands collectively called Montagnard. The "Yards" as we affectionately called them, are the original indigenous inhabitants of Vietnam comprised of many tribes that have lived and thrived in western Vietnam and eastern Cambodia for more than two thousand years in tiny villages scattered primarily throughout the mountain valleys.

The name Montagnard comes from French word meaning "mountaineer" and their language is believed evolved from what is called the Malayo-Polynesian language group and is unlike the common Vietnamese language – Montagnards did not migrate from other parts of Asia like the Vietnamese, but from the south, perhaps Polynesia or the area generally north and east of Australia.

They are primarily farmers who practice what is called "slash-and-burn" agriculture where fields are cleared of smaller trees and plants that are dried in the open air and then burned to prepare the soil for planting. After farming a field for three or four years (or when the soil is believed depleted of nutrients), the land is allowed to lie fallow to regenerate naturally while moving to another field to repeat the process in an ongoing cycle. Montagnards do not typically plow the earth, instead they use dibble sticks to poke holes in the ground for seeds or seedlings, a method that helps prevent topsoil from washing away during the summer rainy season. Montagnard men and their sons and grandsons sometimes hunt small game including rodents and monkeys using wooden crossbows shooting sharpened wood or bamboo bolts (without metal or stone tips) and using thick leaves rather than feathers for stability in flight.

Entering a village, we would set two perimeters, one just outside the village and the other closer in while Doc Rivera tested for diseases and infections which were common. He would also care for the injured and wounded, also tending the elders for their many aches and pains by administering aspirin or other modern medications not found in the jungle. Although they had ancient remedies for many ailments, modern medicine saved many lives. We took them food and clothing and when conditions were considered safe, we played ball or played with the kids, also talking with the adults who were always happy to see us. Language barriers went away with a smile and handshake.

During those trips, the Montagnards sometimes repaid us with small gifts, occasionally a long sip…or two, or three…of their potent rice wine taken through a long grass reed. Sometimes they would offer us a few bananas or a stick of sugar cane or some dried, sweet jerky made from whatever meat they had. We never turned them down because we were guests in their home.

---

[59] *MEDCAP stands for Medical Civil Actions Program.*

I was given a Montagnard crossbow which I later partially disassembled and packaged up in a makeshift cardboard box and mailed it to my Dad, but it never arrived. The crossbows were highly sought-after souvenirs and I have always presumed that somebody spotted it in the mail and decided to keep it for themselves or perhaps thought it was a dangerous weapon not suitable for transport home.

As a sign of friendship, the Montagnards presented many American GIs with hand-formed brass bracelets placed around a wrist or forearm, some crude and others more elaborate.[60] I got mine from a young girl about twelve years old during a MEDCAP to a Montagnard village near Dak To called Dak Tomboe.[61] on about my third visit there.

Montagnard children standing next to bamboo fence near their village (Bob Barrett photo)

During the Vietnam War, the Montagnard people had little use for any of the Vietnamese people, north or south, and regardless of the Vietnam government's political positions – a trend that continues even until the present time. Credible reports suggest that they continue to be mistreated even by the current Vietnamese government which consider them sub-human, often calling them by the derogatory name *moi* which is loosely translated to mean "savage" which is a terribly incorrect assessment of their nature.

To many Montagnards, the Vietnamese language is foreign to them and they prefer to speak their own dialect, also practicing their ancient traditions despite the persecution. The current

---

[60] *In 1966, John Wayne made a USO tour in the Central Highlands where he reportedly fell in love with the Montagnard people, one gifting him a bracelet. That tour helped convince him to co-produce the movie The Green Berets in support of U. S. forces in Vietnam. Look carefully at his wrist in many of his later movies and you will see what is reported to be a Montagnard bracelet. Whether a true story or not, it should be.*

[61] *In his Operational Report for the period April through June 1968, Lieutenant Colonel Akers (who commanded the 6th Battalion, 29th Artillery at the time) mentioned the success of MEDCAPS and the village he spelled Dak Tomboie.*

Communist government is reportedly working to "re-educate" them to accept the common language and the Communist way of life – but don't hold your breath.

A proud and happy Montagnard father holds his small child. (Mike Alesi photo)

Recon/Survey Section team members and other 6th Battalion, 29th Artillery soldiers in the back of a truck headed for a Montagnard village for a MEDCAP. We would get as close as possible and then walk the remaining distance usually less than a half-mile. (Steve Stark photo)

Mike Alesi, in left foreground in the above photo wearing a helmet, regularly carried clothing to the Montagnard children sent him by his Mom and Dad that they collected through their church in Detroit.

In our AO, the most prominent tribal members were *Jarai*, sometimes today called *Brâu* and other *Rơ Măm*, whose population in Kontum Provence including the Dak To area was thriving in the many thousands in the 1960s but in 2009 was estimated by the Vietnamese government, sadly, to have dwindled to under one thousand. Clearly, they are being rubbed out, although some reports suggest they are still fighting the Communists. Even some of the more recent Vietnamese travel brochures refer to parts of Montagnard-inhabited areas as *The Frontier* where additional taxes are collected from visitors wishing to travel into parts of the Central Highlands generally north and west of Pleiku City – our old stomping ground.

One research report written long after the war ended states that United States medical teams treated more than 40 million Vietnamese civilians during the war, spending between 500 million and 750 million 1960s dollars. Winning the "hearts and minds" of the South Vietnamese people was a big part of the U. S. plan.

The Montagnards I spent time with were good and decent people, honest and faithful and full of life. They were sometimes mischievous and many of the young men enjoyed a good prank. Some historians and others have suggested they played both sides against the middle by supporting the VC by night and American forces during the day, but I am unaware of any instances of treachery in our AO. Even so, it is logical to presume that at least some were pressed into service and cooperated for the sake of their family and survival of their village. War is the ugliest of all human endeavors and especially horrifying when civilians are caught in the crossfire and have no way to defend themselves.

The Green Berets (primarily the Fifth Special Forces in our AO) were famously close to the Montagnards and fortified some of their villages or built compounds, then armed and trained the men with surplus M-1 Carbines and other weapons which, to the amazement of many Americans, the Montagnards quickly mastered. They learned to defend their villages and kill VC and NVA in support of U. S. interests while also providing valuable intelligence about enemy troop movement. The Montagnard Tribesmen played a key role in defeating the NVA in 1967 and '68. Some say the Montagnards are natural warriors, even the women.

I respect the Montagnards for their loyalty, courage and tenacity and never met anyone who served in the Central Highlands in 1967 and '68 that did not have great affection for them. Although some have suggested they simply took advantage of American generosity and fought only as mercenaries, I choose to trust my memory.

The Battle of Dak To

By late October and into early November our leaders knew the NVA were massing for an attack for several reasons. First, because fixed wing as well as helicopter aerial observers and LRRP teams had independently reported NVA supply convoys and troops moving from west to east toward Dak To from just inside the South Vietnam border. At the same time, others spotted what appeared to be reinforced bunkers on some of the hills to the south and west of the Dak To airstrip. Also, and probably most importantly, on 2 November 1967, an NVA reconnaissance Sergeant named Vu Hong surrendered to an ARVN outpost and was turned over to the 4th Infantry Division intelligence people at Dak To who instantly transformed the Sergeant into a songbird. He provided a tremendous amount of information that reinforced what was already suspected – at least five regiments of NVA were headed directly toward Dak To while others were massing around Ben Het just to our west while still other units were forming to our south, north and east. Sergeant Vu Hong's information was not perfect, but it was close enough. The NVA were closing in and we were surrounded and outnumbered.

Even before Sergeant Vu Hong turned coat, some U. S. intelligence professionals had suspected the PAVN (NVA) "B-3 Front" was getting ready to move into the Dak To area rather than from further south in the Buon Ma Thout area as others had suspected. This was the reason that we were sent to Dak To.

As it turned out, the 1st NVA Division had joined the NVA 24th and 174th Regiments who were apparently already in the neighborhood having remained there following the fighting earlier in 1967 in that same area.

As of late October, and early November 1967, estimates of enemy strength in our area were as follows:[62]

- 24th NVA Regiment – 1,285 north and east of Dak To.
- 32nd and 66th NVA Regiments – approximately 5,000 southwest of Dak To.
- 174th NVA Regiment – 1,760 west of Dak To.
- 304th LF Battalion (Viet Cong) – 520 south and east of Dak To.
- NVA 40th Artillery Regiment – 1,796 primarily south of Dak To.
- NVA First Division – 2,500 primarily west of Dak To.

Total U. S. manpower in the Dak To area at that time was around 4,500 men compared to the enemy's more than twelve thousand.

---

[62] *Total enemy numbers vary somewhat depending on the information source but are usually within about plus or minus five percent of those reported above that are averages based on the information we found during our NARA visits.*

Instead of waiting for an attack, our leaders developed a plan with three basic objectives in mind: First, to take and control the high ground. Second, close off the enemy resupply routes for men and materiel mainly in the river valleys. Third, destroy the enemy that had been driven from the high ground and prevent him from escaping through the valleys and back across the border into Cambodia.

The First Brigade Commanding Officer, Colonel Richard Johnson (call sign "Cherokee"), immediately ordered his infantry into motion. The first unit to make contact was just one day later on 3 November 1967 when Bravo Company of the 3rd Battalion, 12th Infantry engaged the NVA on Hill 1338[63] under direct artillery support of Bravo Battery, 6th Battalion, 29th Artillery which was then located at Dak To. Others quickly joined the fight including additional First Brigade infantry units and later elements of the 173rd Airborne, all now under operational control of the 4th Infantry Division.

At about the same time, the enemy shelling of Dak To started, first with his 82mm mortars and then using his much more powerful 122mm rockets with their recoilless rifle and artillery fire joining in.

Visible on the left is part of the First Brigade, 4th Infantry Division TOC at Dak To that was a primary target for NVA gunners during November 1967 and several months into 1968 as they repeatedly shelled the area. The men in the right-foreground are part of the Recon/Survey Section that are performing a crater analysis. (Steve Stark photo)

---

[63] *One source says that the contact was made less than one kilometer from where Alpha Company, 2nd Battalion, 503rd Infantry, 173rd Airborne was almost wiped out on 22 June 1967.*

From then on, Charlie did his best to upset any plans we might have by firing on us every day at various times, often late morning, around mid to late-afternoon and usually around two or three each morning just to keep us on our toes. When mortar or rocket attacks came at night, somebody (anybody) could ring the homemade alarms made using an expended 105mm casing (there was no siren at that time) that were hung by a rope or chain and struck with the perforated metal primer removed from the casing.

105mm howitzer casing and striker painted artillery red used as an alarm. (Bob Barrett photo)

When attacks came during the day, the bells seldom sounded, and we would usually just hunker down in place and hope for the best. Getting caught in the open during a mortar or rocket attack was always exciting but the odds of getting hit were not very high and many times the attacks were over before we could even head for cover. Even when our great Radar Section guys spotted rounds on the way in, there was just not enough time. When they did spot incoming, their calculations of the back azimuth[64] were amazingly precise.

Proof that sandbags worked at the Dak To forward base in mid-November 1967 where a recoilless rifle round almost found its mark. (Steve Stark photo)

---

[64] *Calculating the direction from which a shell originates is fairly easy. The trick lies in determining the distance. Both direction (azimuth) and distance (range) are required to determine the point of origin.*

Sometime later a siren was installed near the middle of camp but on most occasions, whether bell or siren, the alarm would not sound until after the first rounds had hit. Many times, however, we did not hear the mortar detonations because Dak To was full of tents of all sizes and shapes and sandbag bunkers which undoubtedly helped baffle and deflect the noise. The 122mm rockets and NVA artillery rounds were another story because they produced a heavier concussion.

While on the south perimeter or in the parapet atop the battalion TOC, we could sometimes hear the sound of mortar rounds leaving their tubes. I never heard them coming in until after they hit although some said they could. By contrast, the sound of a 122mm rocket on its way into Dak To and passing overhead was hard to miss and made a roar like an artillery shell before impact.

In a very short time almost everyone learned to recognize the difference between the sound of incoming and outgoing. To me, the outgoing artillery was more a "ka-boom" while the incoming was more a "bang." Even so, everybody was fooled occasionally and would jump at the sudden noise. Later when the batteries were moved out of Dak To onto the various fire bases, the problem of recognizing the difference went away since there was no outgoing.

One late afternoon during chow, a 122mm rocket or artillery shell slammed into the deep ravine just to our east and detonated with a loud crash. The mess tent was open-sided but screened in to keep bugs out, and at least one of our guys ran completely through the screening without stopping as others charged out through the same hole or through the spring-loaded wooden door to get away. Incidents like that are fun to talk about now but not so funny when it happened. Had the round landed another fifty feet to the west it could have ruined our day.

We were mortared and rocketed many times but had it easy compared to the guys in the hills to our south and southwest who had it much rougher – and we knew it. As the dead and wounded arrived, our battalion Surgeon, Doctor Rivera and his helpers did their part as everybody pitched in to help, also unloading medical supplies arriving from the rear in trucks also carrying fresh troops and ordnance.

Because ground attacks were considered inevitable, everybody at Dak To, regardless of his MOS or normal duties, was assigned a defensive position at one of the three-tiers of roughly concentric circles with the innermost circle being an area around the First Brigade TOC and the nearby helicopter landing pads. The Recon/Survey Section was assigned to help defend the south perimeter near the river where we would join whatever infantry unit was stationed at Dak To at that time.[65] We were very familiar with that part of camp because that is where we also regularly pulled night perimeter duty.

---

[65] *Infantrymen just in from the bush for such duty sometimes called this "Palace Guard Duty."*

Early November 1967 showing the first bunker at Dak To manned by Recon/Survey Section on the south perimeter just north of Dak Poko River about fifty meters. The gathering shown was in preparation for a "mad minute." (Steve Stark photo)

As the fighting developed in the jungle to our south and west, sorties of fighter aircraft made bombing runs on the hills as artillery rained down that eventually turned some of the hilltops into barren treeless moonscapes. The F-4 "Phantom" fighters were especially distinguishable from others thanks to their turned-down horizontal stabilizer and the dark exhaust smoke from their twin jets as they attacked NVA on the ground. Other planes including the F-100 Super Sabre, F-105 Thunderchief, and A-1 Sky Raider delivered bombs of different sizes as well as napalm.

An unknown type of propeller aircraft drops his bombs just outside the west perimeter right across the Dak Poko river at Dak To. Photo taken from the south perimeter. (Steve Stark photo)

Other planes dropped napalm to our west and just across the river. (Glen Farasy photo)

We never saw the giant Air Force B-52s that flew in packs and worked much further away, mainly out to our northwest. The code name for their missions was ARC-LIGHT and the Air Force experimented with many types of new technologies that would eventually lead to the development of smart bombs and other tools of war.

The NVA were deeply entrenched and well protected in elaborate underground bunkers which we would later learn included large sleeping quarters, kitchens, mess areas, hospitals and even recreational areas. They had stockpiles of munitions, food, water and supplies that were replenished on foot using intricate trail systems and, in some ways, they were as well fed and fortified as us and at that time Charlie had the additional strategic advantage of holding the high ground. But we were sure the tremendous firepower would destroy them or at least significantly soften their resolve. We watched in amazement and wondered aloud how anyone on those hills could possibly withstand such a pounding, not realizing how deeply they were dug in.

At night, the *Spooky* fixed-wing ships[66] did their thing, dropping flares and then firing miniguns into the jungle below for close air support as laser-like streams of red tracers shot toward the earth followed by that hollow whirring sound, almost like a low hum. We also watched green tracers streak skyward from the NVA positions and I thought about the old Army adage my Dad learned in Europe during WWII that tracers work both ways.

---

[66] *C-47 transports, akin to the civilian DOUGLAS AC-47VDC-3, had been modified to drop flares and accommodate very high output machineguns called "mini-guns." We called those ships "Spooky" or "Puff the Magic Dragon," sometimes just "Puff" as a spinoff from a popular folk song of 1963.*

A 155mm howitzer raises dust as it pounds one of the nearby hills south and west of Dak To occupied by dug-in NVA. (Steve Stark photo)

A 155mm howitzer in full recoil delivering a shell to Hill 1338 from Dak To in November 1967. (Steve Stark photo)

As the battle plans developed, troops boarded Slicks at the Dak To airstrip and from near Ben Het to be lifted into the hills by helicopter into marshalling areas as commanders on the ground and in the rear continued refining their plans for taking out the NVA.

During preparatory fires into the hills we stood immediately behind 155mm howitzers located only about twenty-five meters south of our Recon/Survey tent. When the light was just right, we could easily see the projectiles as they left the tubes and flew, almost float in slow motion, all the way to their target and detonate with bright flashes followed by smoke and sometimes dirt blasting skyward. The sound of the detonations took only a few seconds to reach back down to us in the valley below.

Over the following days and weeks other artillery batteries of various calibers on other fire bases in the area also pounded the NVA on several hilltops to our south to take out the anti-aircraft guns and mortars intended to support their comrades on other hills.

Glen Farasy remembers that one afternoon while Gary Herder was on BC scope duty, he spotted what he thought was the enemy digging on the north side of the airstrip beyond the perimeter in the low hills. After reporting the incident, the grid was cleared and eight-inch howitzer fire was brought to bear on the target that solved the problem after only a single adjustment. Although other field artillery pieces could deliver more rounds quicker, the eight-inch howitzer was our most precise.

From almost anywhere at Dak To we could see or hear Air Force and Navy jets making their bombing and strafing runs. The napalm cannisters were easily visible as they wobbled to earth before hitting and breaking apart in a bright orange fireball even at midday; then that dark grey-black smoke rising back into the sky. When the wind was right, the fireball was followed by the smell that's not like burning gasoline or diesel but something in between – sort of like you sometimes get while walking out onto the tarmac to board a small plane. And it lingered.

The entire area around Dak To was buzzing constantly with all kinds of aircraft including jets, various transport planes, Bird Dogs, Slicks and Gunships, Hooks, Loches, and Bubbles. Even occasionally we would see what looked like the Corsairs of World War II fame flown by ARVN pilots mostly from affluent Vietnamese families – they worked mostly to our north and east where the ARVN ground troops were also keeping busy.

Wednesday, 15 November 1967 was a particularly busy day that many of us remember very well as documented in my diary on that day:

1600 Hrs. (approx). Another alert was sounded and 82mm mortars began dropping in general area of ammunition dump. Several rounds were direct hits on dump and a fire followed by hundreds of explosions. Several alerts followed throughout the night.

I was one of the men on duty in the parapet on top of the TOC from late morning until around 1600 hours as we received mortar and rocket rounds several different times during the day from the south and southeast. From our perch, we could easily see and hear the transport planes landing on the airstrip, all from west to east. As I was about to go off duty, two of them were stopped near the east end of the runway when one was hit by a mortar round just forward of the wing and immediately began to burn, putting off thick black smoke and flames.

Photo taken from the parapet atop the 6th Battalion, 29th Artillery TOC at Dak To on 15 November 1967 as flames and smoke boil from a C-130 cargo plane hit on the runway. (Steve Stark photo)

At about the same time, rounds began falling in the vicinity of the ASP (ammo dump) and it was finally hit and caught fire just as Sid Hegener and Sergeant Major Charles Taylor were in the vicinity and moving toward one the craters to perform a crater analysis.[67] Spotting a nearby water trailer, they attempted to extinguish the flames but their efforts were not successful as the flames grew higher as some of the small-arms ammunition in the area began to "cook off" due to the heat. For their courage that day, Sergeant Major Taylor earned the Bronze Star and Sid earned the Army Commendation Medal, both with small "V" insignia that signifies they were awarded for valor.

Around 1630 hours, I was relieved from BC scope duty by other Recon/Survey team members and climbed down the parapet ladder and went to our tent a short distance away just as more mortar rounds began impacting. Inside the tent, we were all immediately ordered to grab our gear and

---

[67] *One of the many jobs of the Survey & Recon Section was performing crater analyses usually after, but sometimes during, an attack to locate a crater, quickly study the blast pattern, dig into the crater to locate the fuze or mortar tail fin, take measurements using a line and a compass, then determine the angle of impact and the azimuth from which the round had come to estimate the direction and distance to the enemy position. Technically, this is straightforward since a mortar travels in a predictable arc and so does a 122mm rocket. Recoilless rifles, on the other hand, are direct fire weapons and trickier to figure out.*

head for the south perimeter to take up defensive positions in one of the small, above-ground bunkers.

A rumor spread that an all-out ground attack on Dak To was anticipated which made sense because everybody from the brass on down knew there were plenty of NVA in the area and more coming in all the time. One of the people on the perimeter, an infantryman, was a loud-mouth Texan who joked about how our situation saying it was like Crockett and Bowie and Travis at the Alamo who waited in a similar way in March 1836 as Santa Anna closed in for the kill. We had all grown up with Fess Parker playing David Crockett on TV and knew how that had turned out. I did not think it was funny.

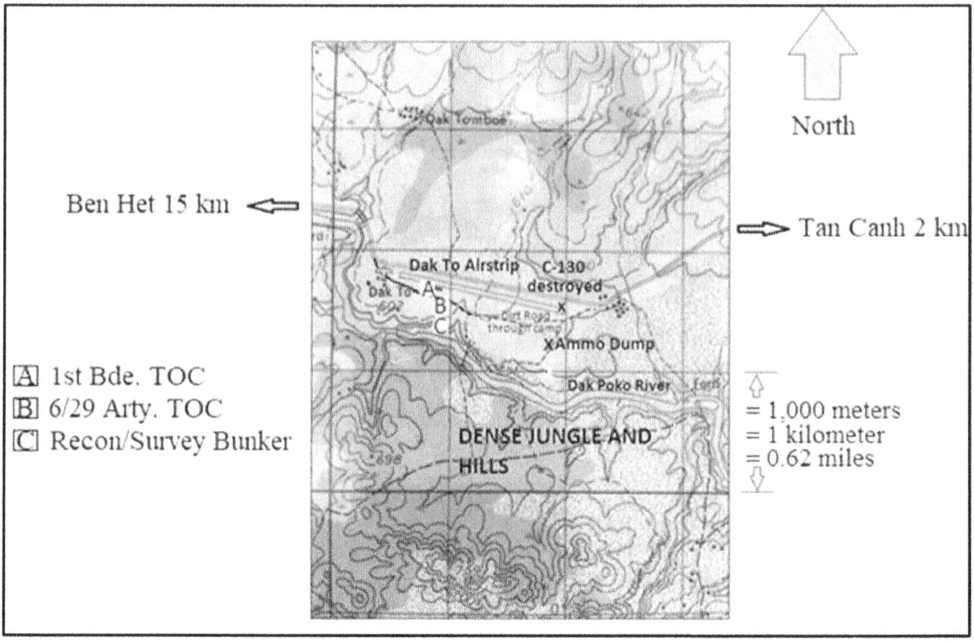

Map of the Dak To forward base area showing the location of the First Brigade TOC, the 6th Battalion, 29th Artillery TOC, the Recon/Survey bunker on the river that was completed in early 1968. Also shown are other points of interest.

From the perimeter, we could clearly see the ammo dump on fire and see incoming rounds hitting mostly to our east and northeast and not in our vicinity. We also clearly heard recoilless rifle fire coming from the hills to our southeast which I have always believed was doing more damage than the mortars. Although some experts have written that it was a "lucky shot" by the mortars that started the ammunition supply fire, I believe the recoilless rifles finished the job by firing directly into the middle of the supply from our southeast and across the river in the low hills.

As larger artillery and other ordinance began cooking off in the ammo dump, the fire grew even larger and many shell fragments fell from the sky in what sometimes looked like a big Independence Day celebration. It still amazes me that none of us were wounded because we had limited overhead protection and many of us were outside in fox holes near the small bunkers.

Looking east toward the ammo dump from the south perimeter at Dak To late in the afternoon of 15 November 1967 as smoke rises from the ammo dump fire. (Steve Stark photo)

That evening as things quieted down, I walked up the hill and returned to our tent for some sleep. Just as I stepped inside our tent, I was hit with a huge concussion, the largest explosion I have ever heard that lifted me completely up and blew me completely over my cot through the mosquito net, tumbling upside down into the cot next to mine and onto the floor. Not knowing what had happened, I caught my breath and looked around in the dark as others also recovered from the blast as someone said, "Is everybody okay?" to which he received all good news. Going outside, the air was full of dust as other stumbled around in wonderment about what just happened. My first thought was that a 122mm rocket had landed on us but there was no smell of explosives in the air and no apparent shrapnel damage to any of the tents. Walking over toward our BC scope parapet, Glen Farasy was on duty but not hurt either despite being higher in the air. He later said he saw the shock wave coming across camp when the ammo dump went up – Charlie had either hit something there that set off the explosion, or something had "cooked off" from the other fires. We were well less than a half mile away from the ammo dump and additional shrapnel began raining down but again, miraculously, nobody was hit. I returned to the river rather than staying in the tent and found cover under some sheet metal and spent the night there.

Everybody who was anywhere near Dak To that night remembers the explosion which has been written about many times, often quoting soldiers who said they thought Charlie had detonated a nuclear bomb. We were later told that a large quantity of C-4 or TNT had gone off and I have heard that our explosion ranked among the largest of the Vietnam War but stories like that are common and you never know for sure what to believe. The most frequently reported time for the explosion was 2145 hours (9:45pm) which sounds about right.

Throughout much of the month of November some writers, journalists and other civilian pests regularly got in the way even as the fighting continued and our dead and wounded arrived at the Dak To helicopter pads. Many later wrote stories as if they were right in the middle of the fight when they actually remained in the rear even at Camp Enari, picking up whatever stories they could get. During this time, a then well-known Associated Press reporter named Peter Arnett and others managed to catch a ride on a fixed-wing transport into Dak To and were standing outside the First Brigade TOC and near the helicopter pad pressing anybody who would talk to them for details. Bob Quinn, the First Brigade Assistant S-3 was there when Arnett asked if our experience at Dak To was going to turn into another Dien Bien Phu – the decisive battle in May of 1954 that culminated with the French being driven from Vietnam. Quinn, a no-nonsense infantry officer, replied, "We are not French and we're kicking their ass" even as the media people continued mercilessly pressing for details and began bullying the Division Public Affairs Officer, a young Lieutenant. Our Division Commander, Major General William R. Peers, emerged from a briefing in the TOC and heard the commotion. Peers, a *Soldier's General*, immediately approached the news people and told them they were not conducting their business in an acceptable manner and ordered the Military Police to escort them to an aircraft and get them out of the 4th Infantry Division Area of Operations. This event was not reported by the media but that did not matter because the Battle of Dak To was clearly an American victory, even finally admitted by Arnett and others who reported that, "…every objective was taken" which was the truth.[68]

General Peers was perfectly suited for his Central Highlands command because during World War II he had fought the Japanese in Southeast Asia with Detachment 101, a clandestine organization under control of the Office of Strategic Services (OSS) forerunner to the CIA. Their intelligence gathering and sabotage greatly aided the 5307th Composite Unit (Provisional), code name "GALAHAD," later popularly known as Merrill's Marauders. Deep in the mountains and jungles of Burma, Peers, then a Colonel, had honed his skills and became an expert in guerrilla warfare

---

[68] *The idea that Dak To would become another Dien Bien Phu was apparently a popular idea among the press as the headline of a United Press International report appearing in several U. S. newspapers including the Palm Springs, California newspaper which on 15 November 1967 printed the headline: "Dak To Devastated by Mortar attacks in Dien Bien Phu-Like Valley."*

and training indigenous personnel – similar to what our Green Berets would accomplish during the Vietnam War. Peers later served with the Central Intelligence Agency (CIA) during the Korean War before taking command of the 4th Infantry division in early 1967.

The Battle of Dak To was the largest battle of the Vietnam War to that time and among the largest of the entire war with both sides terribly bloodied, although enemy losses were greater than ours by about four to one and so heavy that the NVA's 174th, 32nd, and 66th regiments were unable to participate in the coming Tet Offensive just two months later.

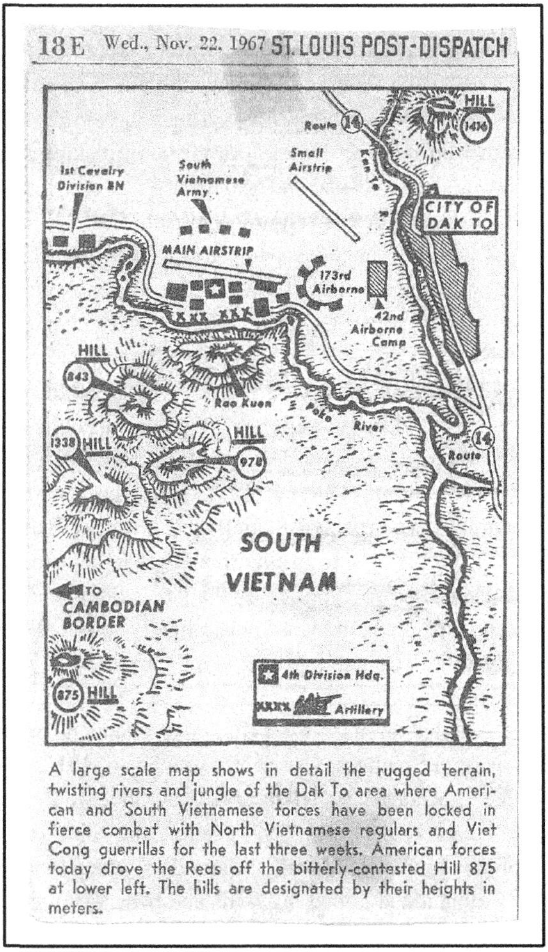

18E Wed., Nov. 22, 1967 ST. LOUIS POST-DISPATCH

A large scale map shows in detail the rugged terrain, twisting rivers and jungle of the Dak To area where American and South Vietnamese forces have been locked in fierce combat with North Vietnamese regulars and Viet Cong guerrillas for the last three weeks. American forces today drove the Reds off the bitterly-contested Hill 875 at lower left. The hills are designated by their heights in meters.

This map of the Dak To area accompanied a long article in the St. Louis Post Dispatch on 22 November 1967. The map is poorly drawn, and the article was ripe with erroneous information. (from scrapbook kept by my Dad and Mom.)

Artillery Liaison Duty with the First Brigade, 4th Infantry Division

One morning probably during the second week of November as the Battle of Dak To developed, I was ordered to report to our battalion headquarters. There, I was told that I had been "selected" for an assignment with the First Brigade as an Artillery Liaison Specialist. They said I would remain based at Dak To and still serve with the 6th Battalion, 29th Artillery as a member of the Recon/Survey team but would spend most of my time either at the First Brigade TOC or be sent to fire bases and other field locations as required. The man I was to replace had just been sent to one of the infantry units as an RTO to replace a man who had been wounded.

Once again and for a third time in less than two months in Vietnam, I was being sent where the Army wanted me. Worse yet, I had absolutely no idea what an "Artillery Liaison Specialist" was supposed to do.

I reported to the First Brigade TOC, a large sandbag-covered fortress of sorts, and found Captain William E. Sammons, the First Brigade LNO, who asked me if I knew the phonetic alphabet, understood how to read a map, and if I knew Army radio protocol. I answered "Yes sir" to the first two and "No sir" to the third. I must have passed his test because he then told me to report to Staff Sergeant King D. Simmons and Specialist 4 John Parent, both on duty at the TOC. At first, Sergeant Simmons was difficult to understand because he spoke with a speech impediment, but I quickly grew used to it and it never caused me any problem. John Parent, a New Yorker, was curt and to the point and knew his business and was a good teacher who, without any fanfare, handed me a pad of paper for notes and immediately began explaining the purpose of the several wall-mounted maps in the very small room of the TOC where the Artillery Liaison Team was located. He then put me to work right away operating the radio to get rid of any jitters, which worked. He put me to work reading the previous several days of the log book and notes he had taken from the most recent intelligence briefings which reinforced the rumors that we were surrounded. A mustachioed Major, who always reminded me of the comedian Ernie Kovacs, came in and spent about two minutes going over the rules regarding confidentiality which was basically just "keep your mouth shut." There were two field phones in our area, both with stickers on them that said, "Charlie is Listening," a reminder that the NVA had people who spoke perfect English just as we had people who spoke perfect Vietnamese.

Captain Sammons' radio call sign was November-Two-Two, Sergeant Simmons was November-Two-Two-Alpha, and mine was November-Two-Two-Charlie. Shortly later, Charles E. ("Charlie") Schumacher joined our team as did Michael A. ("Mike") Wright and the six of us worked very well together.[69]

---

[69] *Near the end of my tour of duty in early September 1968 Roy D. ("David") Graham joined us and was a quick study.*

What nobody told me right away, but which I soon learned, was that my assignment at the Brigade TOC would be mostly during the hours of darkness and beginning around sixteen hundred (4:00pm), ending some unspecified time after sunup. But the good news was that when I got off duty, I could eat a hot breakfast at our mess tent (always the best meal of the day) and then sleep during the morning and early afternoon. I would also still have time to sometimes man the BC scope, pull daytime perimeter guard duty, and other duties "as required." I usually only slept five or six hours a night anyway, so the new duty schedule was no big deal. The downside was that my Recon/Survey buddies would be doing other things while I slept, so I'd miss out on some of the artillery survey work, MEDCAPS, and the other fun duties and projects.

Later that evening after chow I went on duty to begin my new assignment, also meeting others in the TOC including an enlisted Airman who ran a small Air Force radio station set up right next to our liaison area who used the call sign *Tollhouse Lima.* His primary duty was to remain aware of Air Force and Navy air traffic in the area and provide instructions to pilots as to where they should not fly to avoid our artillery. Later, as a bonus, we learned that he always knew where and when the ARC-LIGHTS, air strikes and other Air Force activities would happen. He also had direct radio communication with the FAC pilots who, in addition to their regular duties of coordinating fighter strikes, often provided intelligence about enemy ground activity spotted while performing their regular duties. Tollhouse Lima would pass that information along to us and we would log it, plot it on one of our maps and then pass it along to whatever infantry or other ground units were operating in the area and to the First Brigade planners.

The Artillery Liaison Team at the First Brigade served many purposes including what was called "clearing fires" which was monitoring air and artillery strikes to help ensure targets were far enough from friendly forces to be safe and far enough from civilian areas to avoid collateral damage. To do this, we needed to know which batteries were firing and from what direction in order to make a judgement call for them to proceed – a call made quickly and often without complete information. It was not an exact science.

LRRP

The First Brigade Long Range Reconnaissance Patrol (LRRP)[70] teams were inserted and extracted often by helicopter primarily to observe and report enemy locations and gather intelligence although they sometimes engaged in ambushes. The insertion method sometimes employed multiple helicopters, including one command and control (C&C) helicopter and others that would leapfrog and feign several landings at different locations before inserting a team in order to mask the actual insertion point. The Hueys often would not actually land but instead briefly hover as low as possible as team members stood on landing skids and then jumped to the ground. Occasionally they would repel or "rope-in" when required by the terrain. The typical team in our AO consisted of four or five men, but the actual team size depended on many variables and the intent of the mission. All LRRP team members were volunteers.

One of my jobs at the Frist Brigade was tracking their movement and position which was often challenging because they were so small and agile and able to negotiate even rugged terrain much better than infantry units. It was often difficult to know how far they might have traveled even over a period of just an hour or two. Although the First Brigade protocol required LRRP teams to check in by radio at more frequent intervals than other units on the move, they were sometimes not able to call in depending on their location and other conditions including proximity to the enemy.

LRRP teams traveled light and moved from one location to another with only a compass and map which required the application of skills learned during training as well as on the job. As with all units from time to time, sometimes LRRP teams could not determine their location especially in areas where visible landmarks could not be found on maps.[71] When that happened, artillery batteries would sometimes fire a smoke round or illumination (even sometimes HE) to a known coordinate so the team on the ground could determine their location using the sight or sound of the round as a marker.[72]

LRRP teams also sometimes incorrectly reported their night positions only to discover they were somewhere else when "friendly" harassment and interdiction (H&I) artillery fire impacted too close for comfort. When that happened, radios at the First Brigade TOC would come alive with excited chatter, often including an order of "check fire" to stop the incoming. I am unaware of

---

[70] *LRRP is pronounced "lurp."*

[71] *Like all topographic maps of that era, the maps of Vietnam were often imperfect especially in mountainous regions, dense jungle or other areas drawn by cartographers where scant information existed or was based on outdated maps and often without the aid of land survey data.*

[72] *This was possible because a skilled Battery Commander of a properly positioned battery knew precisely where his artillery pieces were on the map, where they were pointed, and how far a projectile would travel (and impact) when fired.*

any circumstances in our AO where LRRP team members were wounded by friendly artillery fire. Despite the best efforts of everyone involved, men were sometimes lost to errant artillery, air strikes[73] and other mishaps. The term "friendly fire" is a terrible oxymoron.

Several LRRP teams operated in our AO and they undoubtedly saved many lives. The LRRP team RTO that I communicated with most often had the call sign *Hawkeye-One-Charlie*[74] and the team he belonged to seemed always to be on the move. They provided a tremendous amount of valuable intelligence for use by First Brigade planners and others.

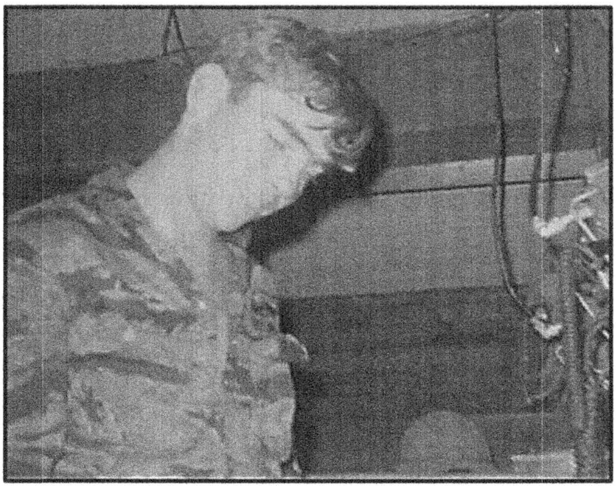

I have long forgotten the name of this LRRP who was wearing the trademark Tiger-Stripe fatigues when the First Brigade Artillery Liaison Team and others met with him at the First Brigade TOC at Dak To in mid-April 1968 after his team returned from three days of reconnoitering the Dak Pek area north of Dak To. (Steve Stark photo)

---

[73] *On 19 November 1967 during the Battle of Dak To, a U. S. Marine pilot flying an F-4 Phantom inadvertently dropped a bomb on the 2nd Battalion, 503rd Infantry, killing 45 U. S. soldiers and wounding another fifty in one of the worst so-called "friendly fire" incidents of the Vietnam War.*

[74] *Based on radio protocol of that time, the "Team" name was probably "Hawkeye One," so the radio call RTO's call sign was most likely Hawkeye-One-Charlie.*

December 1967

After most of the fighting was over in November and into the first part of December 1967, things in the Tri-Border area got eerily quiet and it was back to day-to-day grind of clearing and plotting H&I fires, pulling perimeter guard duty and other mundane chores. By that time, I had been in country for less than three months, but it seemed much longer. Occasional mortar attacks kept us on our toes as the infantry patrols and LRRP teams continued their dangerous work and the Special Forces guys also kept busy. Although there was occasional contact there were no major engagements.

We began hearing about the big national Vietnamese holiday called Tet[75], that is celebrated throughout the country annually at the beginning of the lunar new year. Rumors also circulated that there would be a cease fire around Christmas and maybe lasting even into January and later into Tet and we were told that if a truce or cease fire were called, we would honor it and leave the treachery to the enemy – nobody trusted them. Even so, we would be ready.

Rumors also circulated that Bob Hope and Raquel Welch would be at Pleiku sometime over Christmas or New Year for a big USO show but exactly how people would be selected to attend was a mystery. We all hoped we would be chosen, mainly because Raquel Welch was the hottest thing going at that time, and a very patriotic lady.[76]

In mid-December I received orders to report to Camp Enari for training on what they called sound-ranging equipment, some newfangled device that would supposedly tell us instantly where enemy fire was originating regardless the caliber or type. No matter the reason, I was ready to get out of Dak To and reported to the helicopter pad the following morning where the officer in charge inspected my rifle and then pointed to a Slick headed for Camp Enari and I climbed aboard, muzzle down as always. Besides the crewmembers and me, there were only three others, two MPs, a Staff Sergeant and PFC, and their prisoner – a sharp-looking, well-groomed NVA soldier handcuffed from behind with his knees also bound together with a heavy leather strap. The MP sergeant sat on the left side of the helicopter with the prisoner next to him and me on the right side. The MP PFC sat across from the prisoner. Not a word was spoken the entire flight by anybody on board and I always supposed the prisoner was being taken back to Pleiku for further interrogation, hopefully by people who knew what they were doing.

At Camp Enari, I reported in at the 6th Battalion, 29th Artillery rear HQ and was given instructions as to where to report the following day for my training. I took a warm shower and caught a ride into Pleiku City for a haircut and shave, also getting my boots shined – they had become less than

---

[75] *Tet is short for "Tết Nguyên Đán" that translates to "Feast of the morning of the first day."*
[76] *Glen Farasy was Chairman of our reunion in San Antonio, Texas in 2018 and did his best to contact Ms. Welch to invite her to attend but to no avail.*

presentable at Dak To over the preceding month or so, and at base camp they liked that spit and shine look. It felt good to be clean for a change.

The next day, I reported to a training area and joined up with about a dozen others who were serving with various artillery units been sent there for the same instruction. Our instructors were all American civilians who apparently worked for the company that had developed the device called the Sound Ranging Set AN/TNS-3. We spent two day in classroom training and then two days hands-on in the field practicing what we had learned. On the fifth day, we set the equipment up in an open field while a mortar crew fired 81mm parachute-flare mortars from about a half-mile away so we could test our skills. Later, they fired an M60 machinegun from a lightly wooded area for a similar test. The equipment seemed to work fairly well but it was not much of a test since we could see the mortar crews when they fired and hear the machinegun even without the equipment. All the while during the training, and besides the instructors, there were other observers around us, both Army and civilian always taking notes. They never asked any of us our opinion – just watched us and photographed us and took notes.

The following day I flew back to Dak To and a few days later one of the civilian instructors arrived with two of the sound-ranging units, batteries, and all the accoutrements. We set them both up on the south perimeter just before Christmas in two different bunkers and the instructor left that afternoon.

Instead of seeing Raquel and Bob in Pleiku on Christmas, I got to babysit a sound-ranging device which I tried to use on two occasions during mortar attacks but to no avail. Soon afterwards, both of the units failed to power-up on the very same day and that was the end of that.

Many years later on the Internet I discovered a report dated 22 April 1968 written by the U. S. Army Artillery Board at Fort Sill titled *U. S. Army Test and Evaluation Command Commodity Service Test Procedure, Sound Ranging Set.* The report addressed failures in field testing and made recommendations for further tests. During the Vietnam War, many new technologies came on the scene, some successful and others apparent failures.

Snapshot of the front of one of the sound-ranging set AN/TNS-3 in a bunker on the south perimeter at Dak To in December 1967. (Steve Stark photo)

One of the more successful devices developed during the war was called a "people sniffer" that was intended to "sniff out" enemy soldiers. When I first heard about it, I thought it was a joke and got a big kick out of its nickname that was "Snoopy," but it was for real. It was a chemical sensing device whose "sniffer" could be hung from a Huey flying at treetop level over suspected enemy locations that could detect the presence of ammonia and other gases given off by enemy troops on the ground from perspiration and other bodily functions. One rumor that was never confirmed was that Snoopy could even tell the difference between American and Vietnamese.

One evening in early January I was asked if I would like to take a ride on a *Snoopy* mission as an observer. I eagerly accepted and was shown a map of the area to be covered. When I asked Captain Sammons what I was supposed to do, his answer was something like, "Observe and provide support as needed…" which was typical of many verbal orders which meant, "…figure it out as you go along." I was told to report to the helicopter pad at thirteen-hundred hours (1:00pm) the following day.

The next day (after spending all night at the TOC), I arose late morning, cleaned up, ate, and headed for the helicopter pad. Other "observers" were there as well, and everybody climbed aboard either the Snoopy ship or one of the two support gunships.[77] We took off and flew a wide arc to the west then south and back east in a wide zig-zag pattern with the Snoopy ship at treetop level and the support ships a little higher and behind. I was not wearing a flight helmet, but somebody handed me a headset about half-way through the flight and I listened to the occasional radio chatter which, as I understood it, was all about changing direction, location and speed, all in pilot jargon.

Part of the information gleaned from that flight was used in plotting that night's H&I fires although I do not remember if that work resulted in a body count.

---

[77] *The Snoopy mission I flew on had one "sniffer" ship and two gunships that flew above, behind and on both flanks. In addition, there were C&C helicopters in the area and a Bird Dog in the vicinity, both at a much higher altitude.*

January into February 1968 and the Tet Offensive

Although December was much calmer than November, in January things began heating up again as can be seen from entries in my notebook beginning 5 January and culminating with the first few days of the Tet offensive of 1968.

5 JAN '68 0400 HRS
DAK-TO HIT BY 82MM MORTARS SOUTHWEST OF TOC AND BEING "WALKED" INTO THE PERIMETER. RADAR GOT LOCK ON GRID AND "SPOOKY" CALLED IN FOLLOWED BY AIR FORCE FIGHTER BOMBERS. TOTAL OF APPROXIMATELY 25 ROUNDS.

7 JAN '68 2030 HRS
"WHISKY" SECTOR ADVISOR NEAR OLD DAK TO HIT BY N.V.A. MORTARS. AT 0100 HRS RADAR REPORTED TWO GRIDS FROM WHICH ADDITIONAL MORTARS WERE ORIGINATING, BOTH NEAR VILLAGES – ONE SET OF COORDINATES ALMOST DEAD CENTER OF ONE OF THE VILLAGES.

7 JAN '68 2030 HRS
4 MONTAGNARDS AND 3 VIETNAMESE NEAR DAK TO BASE WOUNDED BY ENEMY SMALL ARMS FIRE.

8 JAN '68 AROUND 0100 HRS
2 MONTAGANRDS AND 3 VIETNAMESE NEAR DAK TO BASE WOUNDED BY ENEMY SMALL ARMS FIRE.

9 JAN '68 2310 HRS
"WHISKEY" SECTOR ADVISOR NEAR OLD DAK TO HIT BY AMALL ARMS FIRE AND REQUESTED SUPPORT FIRE FROM ARVN ARTILLERY UNIT.

10 JAN '68 0255 HRS
KONTUM HIT BY ROCKETS.

10 JAN '68 2100 HRS
LRRP TEAM HAWKEYE ONE CHARLIE REPORTED APPROX. 20 TO 40 ENEMY WITH DOGS WEST OF DAK TO ALONG AN EAST-WEST TRAIL 700 METERS EAST OF CAMBODIAN BORDER. ARTILLERY ENGAGED AND DESTROYED.

12 JAN '68 1900 HRS

1ST BATTALION, 8TH INFANTRY HIT BY RECOILESS RIFLE FIRE TAKING 6 WIA, 5 ARTILLERY 1 INFANTRY.

15 JAN '68 0031 HRS

FSB 13 WEST OF DAK TO HIT BY 48 MORTAR ROUNDS, MANY OUTSIDE PERIMETER.

15 JAN '68 1000 HRS

10 MONTAGNARDS NEAR DAK TO AMBUSHED BY V.C. USING HAND GRENADES, KILLING 4 AND WOUNDING 6.

15 JAN '68 (AFTERNOON)

CONVOY TRAVELING FROM PLEIKU TO DAK TO HIT BY COMMAND DETINATED MINES. 24 NVA KILLED – SEVERAL PRISONERS.

16 JAN '68 2000 HRS

DAK TO SPECIAL FORCES CAMP HIT BY MORTARS. 2100 HRS FSB 12 HIT BY MORTARS.

17 JAN '68 0530 HRS

"WHISKEY" SECTOR ADVISOR HIT BY MORTARS AND SMALL ARMS FIRE AND REQUESTED SUPPORT FIRE FROM ALPHA 5TH BATTALION, 16TH ARTILLERY.

19 JAN '68 0010 HRS.

FSB 12 (BEN HET) UNDER GROUND ATTACK BY 60MM MORTARS, RECOILESS RIFLES, RPG AND B-40 ROCKETS, AND AUTOMATIC WEAPONS. "SPOOKY" RESPONDED. U. S. TANKS ENGAGED NVA AT CLOSE RANGE. 105mm BEHIVE EMPLOYED. CAPTAIN SAMMONS ON LOCATION ADJUSTING FIRE. ARVN SOLDIERS PERFORM POORLY INCLUDING FIRING ILLUMINATION MORTARS OVER U. S. POSITIONS RATHER THAN OVER NVA POSITIONS THUS EXPOSING U. S. TROOPS. 20 JAN '68. NVA DIRECT HIT BRAVO BATTERY, 6TH BATTALION, 29TH ARTILLERY HOWITZER CREW RESULTING IN 1 KIA (ROBERT BUTLER) AND 4 WIA. AT 0259 HRS. ANOTHER DIRECT ASSAULT. 0403 HRS. LAST SHOTS FIRED AND DUSTOFF IN FOR KILLED AND WOUNDED. LATER LEARNED THAT FSB 25 HIT AT SAME TIME. ALSO LATER IN DAY PLEIKU CITY AND CAMP ENARI HIT BY MORTARS.

23 JAN '68

FSB 12 HIT BY MORTARS OFF AND ON ALL DAY.

24 JAN '68 2120 HRS

INTEL REPORT THAT NVA 40TH ROCKET REGEMENT NEAR DAKTO.
25 JAN '68 EARLY MORNING
INTELLIGENCE REPORTS SAY PLEIKU WILL BE HIT EITHER EARLY THIS MORNING OR TOMORROW NIGHT.

25 JAN '68 1400 HRS
SNOOPY READINGS SOUTH AND WEST OF DAK TO VERY HEAVY.
26 JAN '68 0340 HRS
CAMP HOLLOWAY NEAR PLEIKU UNDER GROUND ATTACK. ALSO HIT BY MORTARS & ROCKETS.

26 JAN '68
INTEL REPORT THAT NORTH KOREANS HAVE CAPTURED U.S. SHIP.

27 JAN '68 0030 HRS
PLEIKU (CAMP HOLLOWAY) UNDER ATTACK BY ROCKETS, MORTARS, AND BY GROUND.

27 JAN '68 1800 HRS
FSB 29 HIT BY MORTARS. 15 TO 18 ROUNDS, NONE INSIDE PERIMETER – NO CASUALTIES – NO DAMAGE.

30 JAN '68 TET 0001 HRS TET (YEAR OF THE MONKEY) IS HERE.
30 JAN '68 GRID 926265 KNOWN LOCATION OF 5 TO 10 NVA AND/OR VC WITH 200 POUNDS OF T.N.T. WITH MISSION: TO DESTROY BRIDGE AT FSB 3 NEAR DAK-TO. ILLUMINATION TO BE PUT UP OVER BRIDGE BUT THEY CANNOT BE ATTACKED BECAUSE OF TRUCE. SAPPERS MOVING IN ON BRIDGE WERE SPECIALLY TRAINED IN NORTH VIETNAM ON DEMOLITIONS ACCORDING TO CHU HOI (TURNCOAT) WHO WAS WITH THEM. 0100 – DAK TO AWAITING LAND, ROCKET, MORTAR ATTACK. ESTIMATED 2 BATTALIONS, ONE TO OUR NORTH ONE TO OUR SOUTH. DAK TO DEFENSES ON PERIMETER. 0223 HRS. PLEIKU BEING HIT BY MORTARS. 0224 HRS. KONTUM BEING HIT BY MORTARS, ROCKETS, AND GROUND ASSAULT. 0321 HRS. SMALL ARMS BECEIVED RECEIVED EAST END OF DAK TO RUNWAY. 0330 HRS. TAN CAN BEING HIT VERY HARD WITH NVA AND VC BURNING NEARBY MONTAGNARD VILLAGES. ATTACKS IN GENERAL AREA OF CENTRAL HIGHLANDS INCLUDE PLEIKU, DAK TO, KON TUM, TAN CANH, QUINOHN, NATRANG, DANANG, BAN ME THOUT. IN PLEIKU AREA NVA AND VC HIT CITY OF PLEIKU DOWNTOWN, POW CAMP, CAMP HOLLOWAY, CITY JAIL, MONTAGNARD TRAINING CAMP, AFRS (RADIO STATION). DURING ATTACKS, MANY NVA AND/OR VC TURNED THEMSELVES IN IN GROUPS OF 3 TO 5.

31 JAN '68
DAK TO HIT SEVERAL TIMES BY MORTARS, ROCKETS AND RECOILESS RIFLES WITH LIGHT US CASUALTIES BUT HEAVY NVA/VC CASUALTIES. 2100 HRS. INTEL REPORTS NVA BUILDING FIRES 500 METERS FROM THEIR TRUE POSITION TO DIVERT ARTILLERY FIRE

1 FEB '68
0010 HRS – KONTUM REPORTS UNDER GROUND ATTACK. ALSO RECEIVING ROCKETS, MORTARS AND HEAVY MACHINE GUN FIRE. REPORTED TO AIR ADVISOR AT 1ST BRIGADE HQ BY PILOT OVER KONTUM. AT 0240 HRS, TAN CANH HEAVILY DAMAGED PER MP REPORT. CAN SEE LARGE EXPLOSIONS IN TAN CANH FROM FIRST BRIGADE TOC. 0330 HRS – FIRST BRIGADE CO ORDERED TOWN OF TAN CANH HIT BY ARTILLERY TO BLOCK NVA ESCAPE. 0400 HRS NVA CHATTER PICKED UP ON OUR RADIO WHICH WE RECORDED FOR S-3.

2 FEB '68
MORTARS RAINED DOWN ON DAK TO ALL DAY LONG AS NVA GUNNERS OBVIOUSLY WERE FIRING ON HELICOPTER PADS AND THE AMMO DUMP. WE UNCOVERED 82MM TAIL FINS NEAR RUNWAY DURING CRATER ANALYSES.

3 FEB. '68 – DAK TO, R.V.N. 1100 HRS.
MORTARS COMING IN VERY CLOSE.
3 FEB '68 1400 HRS, 1600 HRS, 1800 HRS.
RECOILESS RIFLES, MORTARS, ROCKETS FIRED. 3 KIA 7 WIA. – 1/8 INF. ROCKETS (122MM) COMING IN. CRATER ON RUNWAY 10 FT. DIA – 5 FT DEEP. GROUND ATTACK ANTICIPATED.

4 FEB '68 PREPARATIONS BEING MADE.

During the events depicted above on 19 January, I was on duty at the First Brigade TOC performing my Artillery Liaison Duties and communicating with Captain Sammons when the attack began. I talked with him throughout the night and into early morning to help him track and coordinate his many requests for fire support.

Captain Sammons earned the Silver Star for his valiant actions that long night, rightfully credited for skillfully adjusting our artillery, mortars, and Spooky ships to successfully defend Ben Het.

Captain William E. Sammons outside First Brigade TOC at Dak To in January 1968 two days after the action at Ben Het during which he earned the Silver Star. (Steve Stark photo)

What is commonly called the 1968 Tet offensive took place all over South Vietnam including in the Central Highlands. Although Tet officially began the end of January, we were hit much earlier and attacks in our AO lasted until at least the end of March and probably a bit longer. Although we did not know it at the time, the very tenacious NVA and VC were suffering major defeats all over Vietnam during Tet despite what much of the American press and others were reporting to the world.

Early during the offensive on 3 February 1968 Lieutenant Larry Skoglund was working his way back from R&R in Australia aboard a C-123 cargo plane on the final leg of his return to the war. His assignment at Dak To included working as an aerial observation FO and serving as the Recon/Survey Officer. He had previously been assigned as an FO with the 3rd Battalion, 8th Infantry during the fierce fighting the previous November in the jungles and hills of the Tri-Border Area during the Battle of Dak To. Just coming from R&R, Skoglund did not realize the extent of the Tet Offensive which had begun just a few days earlier while he was still enjoying the well-known hospitality of the Australian people.

Following protocol, he had planned to report to the 6th Battalion, 29th Artillery TOC upon landing but upon landing he was immediately approached by the First Brigade Air Liaison Officer – a Captain, or perhaps a Major – a name now forgotten.

"Larry, I need you right away" were the words Skoglund remembers the Liaison Officer saying as he quickly explained that an infantry platoon five or six kilometers (roughly 3 to 4 miles) west of the Dak To airstrip had run into trouble and had no forward observer and was unable to make any communication with any artillery unit. Furthermore, he said, Skoglund was the only available FO who also had aerial-observation experience.

Still in short-sleeve khakis, Skoglund's first thought was to retrieve his weapon, flak vest, and flight helmet but there was "No time for any of that" said the Liaison Officer. "These guys are in real trouble." Also, the only OH-23G "Bubble" helicopter available had been red-tagged due to electrical issues and both of its on-board radios were out of commission but there were two PRC-25 radios already on board, one each set to separate artillery and infantry frequencies.
Although the mission was not officially authorized, Skoglund had previously been in tight spots and knew what it was like to operate without artillery support, so he made the decision to act.

Without a helmet but wearing a borrowed flak vest and borrowing an M16 and bandoleer of magazines, Skoglund climbed into the tiny ship, checked the radios and nodded to the pilot, Chief Warrant Officer Orrin Leonard Dyer, Jr., whom Skoglund had flown with many times before – and they were a solid team as they lifted off and flew west in the 305-horsepower observation bird toward the action. Using the two radios, Skoglund contacted the artillery and then the troops on the ground.

In retrospect now more than fifty years later, Skoglund remembers that they were flying slowly as they approached the location and as Mr. Dyer started to swing the aircraft around. Suddenly, Skoglund spotted the NVA on the ground and reached over to tap Dyer with his right hand to let him know but both men were immediately hit by small arms fire. Skoglund's right arm was shattered right below the elbow tearing the artery and as he looked over toward Orrin Dyer to let him know he'd been hit, he saw Dyer slumped down and unconscious, never to awaken again.

Some men would have panicked, but Skoglund did not. Instead, he went into survival mode thanks to his previous combat experience.
Larry's own words tell the story best: "Just as cool as can be, I locked the stick between my knees, reached across with my left hand and stabilized the stick on my right, tinkered with pedals for just a second to get us straight and aimed back toward the airstrip. It wasn't pretty because I had to subtly experiment to see what happened with very slight movements of the controls and make corrections, mostly up and down – up and down. Company B of 3/8th Infantry was on a knoll near the western end of the runway and I guess they were all watching. They'd heard the rifle fire and for all I know the patrol under attack was theirs. At any rate, Dean R. Plager, a platoon leader I had known from my FO days later told me that the ship was really jerking all over the place and then right at the end of the runway it went straight up into a loop and then hit the ground and exploded. I remember flying toward the approaching runway and thinking I could set down gently and then immediately switch off the magneto – a precaution that our pilots had drummed into my head. However, because I couldn't hold up my arm to slow the blood loss, the shattered artery had run out of blood. I apparently went into shock as things got fuzzy and I passed out. Clear as a bell, I recall saying aloud, 'fuck it, this is too much work' and letting go of the control in my left hand. My next memory is a vague one; a little consciousness at the medivac pad at Dak To. Then

nothing until I woke up at the 71st Evac at Camp Holloway just north of the provincial city of Pleiku."

The official after-action report of the incident concluded that the helicopter was hit twice by small-arms fire from automatic weapons of 7.62mm caliber as it flew at 700 feet elevation above the terrain at a speed of 70 knots (about 80 MPH), but there is no telling where that information originated. Larry remembers very clearly that Orrin was piloting his ship near treetop level, typically a very safe place since it was possible to fly past the enemy before he heard you approaching, thus offering him less time to react and fire.

The following day, Sunday, 4 February 1968, the 71st Evacuation Hospital at Camp Holloway near Pleiku was rocketed heavily and the perimeter was breached at the Camp Holloway airstrip. The 1968 Tet offensive was in full swing as many other locations around Pleiku including Camp Enari were under intense ground attacks by NVA and VC.

As the enemy ordnance poured in, some shells detonated in the hospital wing where Skoglund lay tied down and with multiple IVs running. Neither the detonations nor the small-arms rounds piercing the thin walls bothered him much – he was in "never-never land" thanks to the medication used to deaden his pain.

When a 122mm Chicom rocket exploded in the wing, there was an immediate response by the 71st Evac staff as they began tending to their patients and other wounded. Even though enemy shells were still exploding inside the compound and small arms fire continued to pop through the thin walls, they removed some patients to safer areas while continuing to treat the wounded. Even in the chaos they remained calm and continued instilling confidence in their patients assuring them that everything would be okay. Skoglund could not be moved due to the IV's and floor stands and due to his burns. He recalls that two nurses came up to him, told him not to worry, and proceeded to take mattresses from adjoining beds to place them on top of him for protection. He remembers that all of them were unbelievably calm and collected and remained very professional. "They protected me as best they could and then continued working with patients in the other damaged areas. I was medicated well, as I felt no panic or fear at being so helpless."

Convincing Larry to let me write down and publish this part of his experiences in Vietnam has taken many years of coaxing, but worthwhile not because any of us needs to hear yet another Vietnam war story but because his history is part of ours and it needs to be told.

Very active with veterans' groups and having taught and worked with young people for many years, he has become an accomplished speaker on the Vietnam War and is well respected by all who know him.

First Lieutenant Larry Skoglund at Dak To (Larry Skoglund photo)

Going Underground at Dak To

After the mortar and rocket attacks in November and December the decision was made that the 6th Battalion, 29th Artillery at Dak To would move completely underground and work began as bulldozers operated by 4th Infantry Division engineers dug trenches about eight feet deep and of various lengths depending on their intended use. They dug what would become the Recon/Survey bunker on the south perimeter not far downhill from our original defensive bunkers. They made the cut about thirty-five to forty feet long running east and west parallel to the Dak Poko River which we laid over north to south with felled trees that were then covered with plastic PSP and plastic sheeting, then several layers of sandbags with plastic sheeting between each layer. It eventually consisted of three chambers, rooms of about equal size with a single passageway running the entire length on the south side and an escape hatch on the west end. It included an elaborate drainage system designed by Gary Herder (who was a proud Purdue "Boilermaker" graduate engineer) and ventilation pipes made from powder cannisters. The main entrance was built on the southeast end facing south with a rough-formed wooden door and an L-shaped blast wall immediately outside. At the same time, a partially covered fighting bunker for guard duty was built just outside the entrance to replace the old guard bunker.

From left to right: Mike Alesi, Ron Holloway, Glen Farasy, and Sid Hegener work on interior. The ends are still open with part of south wall already up and one of the room patricians installed. (Glen Farasy photo)

In our section of the bunker at Dak To, Mike Alesi was the Head Chef and would occasionally prepare a gourmet meal from C-rations or food we borrowed from the mess tent. I have always remembered my favorite being a sandwich Mike made using ham or beef, canned cheese, peanut butter, a little jelly, and salt and pepper. He would start by lighting the little stove made from an

Mike Alesi waves while standing just east of the recently completed Recon/Survey bunker on the Dak Poko River at Dak To in January 1968. Note the power pole (we had electricity!) and a tarpaulin laid down for extra protection during the upcoming monsoons. (Steve Stark photo)

We built this wooden sign to mark the "Survey Cellar" next to the Recon/Survey bunker on the extreme south perimeter at Dak To, also giving it the imaginary address of 1369 South Perimeter Drive. The four pipe-smoking gentlemen are from left to right: Sid Hegener, Glen Farasy, Al Roden, and Mike Alesi. (Glen Farasy photo)

empty C-ration can to simmer the concoction over a fire using whatever fuel he could scrounge up, sometimes just scraps of cardboard from a C-ration or other container. As the main ingredients cooled, he toasted C-ration bread over the fire then hollowed it out to then spoon in the hot mixture to make a sort of sloppy joe sandwich which was warm and always delicious. Sharing my recollections with Mike about his recipe during a recent reunion in San Antonio, he corrected me saying that his sandwiches were made of C-ration ham, pineapple jam, and toasted bread and he does not remember the peanut butter or cheese. I now yield to his memory over mine because you never argue with a great Italian chef! The meals Mike prepared for us were the best I ever tasted while in Vietnam. Thanks buddy!

Mike Alesi chomps into one of his great "sandwiches" in our bunker at Dak To probably mid-January 1968 (Steve Stark photo)

Everybody in Vietnam made the little stoves from empty C-ration cans and many soldiers had their own special recipes that frequently included McIlhenny's Tabasco sauce to enhance the taste (or sometimes to cover it up). The McIlhenny Company of Avery Island, Louisiana provided small bottles of their Tabasco sauce and little cookbooks with recipes for preparing C-rations.[78]

Our bunker, like many other sleeping quarters in Vietnam, was also occupied by large rats and dealing with them was a constant problem. Somebody said that rats couldn't burp and giving them beer would kill them because the effervescence would fatally bloat them. One night somebody poured a half can of beer into an empty C-ration can and placed it on the floor in the open area between the four bunks[79] where Gary Herder, Mike Alesi, Glen Farasy, and I slept while at Dak To. I don't remember killing any rats that way, but one night during the wee hours somebody fired his M16 on full-automatic at a rat on the floor that sent dust and wood splinters flying everywhere and really livened things up. Nobody today remembers exactly who fired those shots except maybe Glen Farasy, but he will never tell.

---

[78] *There were thousands of American companies and other organizations and individuals that stood firmly behind U. S. troops during the war and we owe them all a debt of gratitude.*

[79] *Our beds in the bunker were actually bunkbeds with wooden cots used for the lower bunks and wooden frames for the uppers.*

The following article was printed in Pacific *Stars and Stripes* newspaper, Vol. 24, No. 135, Thursday, 16 May 1968 on page 7, and titled "*Artillerymen Go Underground; No Gripes About Digging*" and reads as transcribed below.

Din of War Fails to Shake Saigon Refugees

Artillerymen Go Underground; No Gripes About Digging

Road Paves Way To Civic Betterment

Chapel Opened At Phu Bai Base

Bad Boy for the Reds

Marines Aid Viet Rice Crop

The article reads as follows:

DAK TO, Vietnam (IO) — If you wanted to make a bad pun you could say the men of the 6th Bn., 29th Arty., are really "digging" their work these days. In fact, the 4th Inf. Div. troops are digging so much now that soon most of the forward command element of the artillery battalion at Dak To will be underground. "About the only things that will be housed above ground will be our mess hall area and the battalion motor pool workshop," said Lt. Col. Albert Akers of Gallatin, Tenn., Commanding Officer. "With everything underground in deep, secure bunkers, rocket and mortar attacks won't affect us like they have in the past." Previously, enemy attacks on the 1st Brigade command camp have sent the troops scurrying for their protective bunkers. "Now, if there's an attack during the night, everyone will be sleeping underground and safe," Akers added. Currently, 95 per cent of the battalion is dug in, according to Akers. Enlisted men, senior NCOs and officer billets are all underground as is the command center and, eventually, the battalion will have deep bunkers for various clubs, the ammunition section and communication and wiremen. The battalion's Survey Section has already completed its living quarters, a well-designed bunker which houses 12 men. "It took us about 15 days to build this one," explained Spec. 5 Sid Hegener of Maredosia[80], Ill. "Although it's dug into a slope, we haven't had any problems with rain or mud getting in. It's kept real dry." The deep, but narrow, bunker was constructed with the usual complement of logs, ammo crates, steel planking and sandbags. "We have a kind of basement," said Spec. 4 Gary Herder of Indianapolis, "and if any water does leak in, it will run into the basement sump." The surveymen have also outfitted their living quarters with as many comforts as their imaginations can provide. Shelves, desks, drawers and a stereo tape system[81] are all built into the bunker. "We've even given the place an address," continued Spec, 5 Hegener. "It's 669 S. Perimeter Drive[82], Dak To, Vietnam. We have a nice place here. It looks out over the river." The 299th Engineer Bn. (Combat) and the division's 4th Engineers have worked closely with the artillerymen, bulldozing out deep holes and displacing tons of earth for drainage ditches and sumps. "We weren't too sure if all the bunkers would stay dry during the monsoons but so far we haven't had too many problems," said Spec. 4 Thomas D. Snare of Fort Worth, Tex., who works in the battalion's Motor Pool Section. "It was definitely a lot of work," added Spec. 4 Randall H. Smith of Las Vegas, "but I guess when we hear incoming mortars or rockets, we won't have to crawl out of the sack and run through the mud to a bunker. We'll already be here."

---

[80] *Meredosia is the correct spelling of Sid's hometown.*

[81] *In fact, the only "furniture" we had was what we built using disassembled pine ammo boxes and straightened nails. Also, there was no "stereo tape system" although some of us had cassette tape players. We believe that Stars and Stripes may occasionally have printed misinformation for Charlie's benefit in the never-ending PSYOPS war.*

[82] *The fictitious address we came up was actually 1369 South Perimeter Drive so the correspondent's notes must have been wrong. Many of us remember the address even today, more than a half-century after we were there.*

March 1968

The entire first half of March in 1968 was a very busy time for us and especially from around the 9th to the about the 15th. On 13 March, for example, the Dak To compound was pounded a little heavier than usual most of the day by enemy rockets and mortars. The attacks became even heavier by mid-afternoon to include recoilless rifle fire.

I had been on duty at the First Brigade TOC the previous night (12 March into 13 March) until around sunup when I ate a light breakfast in the mess tent around seven then headed for our underground bunker for some sleep. Around two in the afternoon I awakened to the thump of incoming rounds and got up to look outside, but I was unable to tell where they were impacting but knew it was inside our perimeter to my north and east. I could also hear our artillery impacting in the lower hills to our south and could see some smoke, so it seemed logical that somebody was firing counter-mortar fire at suspected enemy locations and perhaps at the guys who were shooting at us.

I was not scheduled to go back to the First Brigade TOC until sixteen hundred hours (four o'clock) but decided to get ready and go up early since so much was going on and I wanted to catch up on the day's activities – something we did when things were hot.

Grabbing a quick snack[83] from my footlocker, I gathered my gear and hurried out of the bunker and stopped to listen for more incoming – all was quiet – so I headed up from the river and trotted toward the Brigade TOC, but my timing was terrible. About halfway up the hill I started hearing mortars landing ahead and to my right and it sounded like they were walking in closer but then a couple hit farther away. Hesitating briefly, I started to go back down the hill to our bunker but decided they were hitting far enough away that I could make it to the TOC before they caught me. So, I kept to a steady low trot on up the hill almost to the road.

Now I hear even more mortar rounds hitting, this time even closer and I suddenly feel a concussion of a larger blast, this time to my left that was, I thought, a 122mm rocket this time. I stop momentarily near our battalion TOC parapet next to the ladder and hug the sandbag wall. Almost immediately the rounds stop landing, so I again start toward the First Brigade TOC and cross the small dirt road and spot two guys in the ditch on the north side of the road with one practically laying on top of the others as they both look up at me. I can see the open wound on the right upper chest and throat area of the guy on the bottom and the guy on top is pressing his fist onto the wound

---

[83] *Many families sent what were called "care packages" from home on a regular basis that included mostly snack foods and other small items to make life a little more pleasant. Most everybody shared what they received that was usually kept in little wooden lockers we built from disassembled artillery and mortar ammo boxes that were sometimes painted or otherwise personalized.*

to slow the bleeding. The man on top says calmly, "Can you get a Medic?" and I distinctly remember answering, "Yes, right now," and immediately ran for the door of the Brigade TOC, now only about twenty-five meters away.

Entering the TOC past the sentry, I went to a field phone and cranked it, handed it to somebody who was nearby and told him, "Get a medic to the road outside the TOC." Then, I set my rifle down and unbuckled my pistol belt and headed back outside toward the ditch to see if I could help, only to watch several others carrying the wounded man away.

Throughout the evening and into the very dark night more rockets and mortars pelted the area and especially in the vicinity of our TOC. The thick outer walls and reinforced sandbag roof protected us from harm as several mortars impacted, dust sometimes sifting down through the board ceiling. Well after midnight, one of the volleys blew down our radio antenna and Dale Garrison, our Commo Section Chief, and I climbed to the roof and he repaired it under fire while I held the flashlight. Garrison was calm as could be – he was always cool as a cucumber. Later that night they hit the antennas again, this time an officer in the TOC stopped us from climbing up to repair the damage.

Dale Garrison, 6th Battalion, 29th Artillery Commo Section Sergeant at Dak To in early December 1967 (Steve Stark photo)

Just two days later on 15 March we again came under attack by mortars and rockets that began late morning and lasted until well after dark. During one of the attacks that night, several men just outside the TOC were wounded including a Master Sergeant who came stumbling into the TOC and rolled to the floor facing up but not complaining of any pain. We propped his head on a fanny pack and opened up his fatigue shirt to find a small pink hole in his abdomen about the size of a marble, but no blood. We kept him resting calmly on the floor smoking a cigarette until a medic arrived and quickly began checking him over, then pressed on the area of the wound which caused it to ooze. The Sergeant looked down, saw the blood and his eyes rolled back in his head and he passed out cold. Others arrived with a stretcher and carried him off to Doc Rivera's medical tent. The following day I was told the wounded man in the ditch was an MP who had died, but none of the Vietnam records I can find confirms it. I have no idea the identity of the Master Sergeant or several others who were wounded that night at Dak To.

By that time, everybody – Generals on down to the lowliest Private – understood the NVA strategy in the Central Highlands:

- First, move troops and materiel into South Vietnam from Cambodia and Laos in large numbers and join units already in place which had been digging in, collecting intelligence and conducting probes to test U. S. strength.

- Second, wait until the time and the weather was right, then hit strategic locations and inflict as much damage as possible with help from the Viet Cong.

- Third, as U. S. strength prevailed, pull most of the troops back across the border into Cambodia and Laos where US forces were forbidden to follow,[84] along the way leaving some units behind hidden in the hills to continue harassing US fire bases and to gather additional intelligence and prepare for the next push.

- Fourth, with the main NVA force safe inside Cambodia and Laos, replenish and resupply with fresh men and material through the Ho Chi Minh Trail and train the newly arrived troops while continuing to plan for the next offensive.

- Fifth, repeat the process.

The now crystal-clear pattern repeated itself several times beginning in November 1965 when the Air Cavalry under Hal Moore first engaged him in the Ia Drang Valley, and again just east of that same area in May of 1967 during the first major engagements of the 4th Infantry Division in Vietnam. Elements of the 173rd Airborne again engaged the NVA a third time in the summer of 1967 west of Dak To but did not fare well. Then the fourth time in just two years during the Battle of Dak To in November 1967. The Tet Offensive of 1968 was a little different but once again the NVA attacked parts of the Central Highlands in force including Dak To and Tan Canh. Some historians say that the Battle of Dak To was designed to pull American forces away from the major

---

[84] *In 1970, long after I returned stateside, 4th Infantry Division and other units were finally permitted to "legally" enter Cambodia in pursuit of the enemy.*

cities in preparation of the Tet offensive, but that strategy failed because the NVA were decimated at Dak To which lead to a miserable outcome for the north during Tet in 1968.

Eagle's Nest

While assigned to Artillery Liaison duty we were occasionally invited to sit in on the early-morning briefings with around twenty officers and senior NCOs attending. Our orders were to listen, learn, and speak only if asked a direct question. Most of the briefings were about the same and always included recent intelligence our planning for the upcoming days. Those who gave the reports always included an accounting of wounded and killed on both sides. Sometimes Special Forces Officers would attend and occasionally there were Air Force FAC pilots, Bird Dog pilots, LRRP NCOs and even strangers we had never before seen, some in uniforms with no insignia, unit or rank. It was the "who's who" of the First Brigade and always interesting.

In late April 1968, U. S. intelligence units reported increased enemy activity and road construction in the area immediately north of Dak Pek and it was decided to expand the First Brigade operational area to include the Dak Pek area. On 25 April, a Brigade Task Force (named "Task Force Cherokee") which consisted of Special Forces elements and a small command element from the First Brigade was, according to Colonel Akers' report, "*...formed at Dak Pek and given the mission of screening enemy movement and coordinating the defense of Dak Pek.*"

During a morning briefing about a week later, somebody reported that the Dak Pek area was beginning to heat up according to intelligence reports pieced together from several sources including aerial observation and Snoopy readings. It seemed clear that the North Vietnamese were again going to make a move eastward, this time from Laos into South Vietnam coming across the border just north of Dak Pek where an existing east-west trail – an extension of the Ho Chi Minh Trail – was being improved. One report from an aerial observation plane even said the NVA were using bulldozers. It was clear that the NVA planned to once again follow their never-ending strategy to push into the Central Highlands and move all the way across South Vietnam along Route 19 to the coast.[85]

That afternoon I was asked if I wanted to take a "little trip up north" to see the countryside near the Special Forces camp at Dak Pek which was in a valley on the Dak Poko River and about four miles east of the Laotian border. Questions like that were always answered in the affirmative. Besides, I was ready to get out of Dak To for a while.

The following morning, I was briefed, given maps, and all the radio frequencies in use in the Dak Pek area. I was told to provide whatever support was needed. That was it.

---

[85] *A maxim about Vietnam borrowed from Hal Moore's book "We Were Soldiers Once, and Young" states, "Whoever controls Route 19 controls the Central Highlands, and whoever controls the Central Highlands controls all of Vietnam."*

I gathered my gear, tightly rolled and belted my sleeping bag and headed to the helicopter pad near the southwest end of the Dak To airstrip just northeast of the First Brigade TOC. I checked in and then boarded the awaiting Slick with four or five others as everybody settled into the webbed seating as the chopper lifted off and headed north from Dak To, flying about thirty-five miles to a tiny fire base called Eagle's Nest[86] which was on a long and narrow hilltop running generally north and south.

The small landing pad on the southern end of the fire base was much narrower than the rest of the hill. On my TOPO map, the contour lines reminded me of a T-bone steak – something I had not eaten for a long time. The landing pad was covered with partially filled and flattened cloth sandbags which were hard as rock and looked like big cobblestones. Immediately after the skids touched down everybody on board quickly exited as some of us tossed our packs out the doors as the pilot kept the engine running. At almost the same time, others waiting to get on threw their gear aboard and climbed in. In just a couple of minutes the Huey was again airborne and soon out of sight. Our helicopter pilots knew that they were always a prime target and did not like to hang around any longer than necessary. The courage of our chopper pilots and crews is well documented and everybody admired their tenacity.

Between the landing pad and the top of the hill was a narrow path leading first down into a deep depression then rising again about fifty feet away. This meant that while walking to and from the landing pad, our heads would automatically be much lower than the main rotor blade – something I always thought about, but especially when around the smaller choppers like OH-23s whose main rotors were closer to the ground. Most everybody naturally bent over when boarding and exiting Hueys although the main rotors were plenty high at least on flat terrain. Human instinct is to keep away from even slow-spinning metal blades above your head.

Walking up the hill I easily found the headquarters TOC which was fashioned from a connex where there were three men including a Captain, the highest-ranking officer there, so I reported to him. He welcomed me and asked a Spec 4 to find a place for me to make my nest.

The top of Eagle's Nest was rounded and small, so all living quarters were dug into the hillside at various distances from the crest and spaced perhaps ten to thirty feet apart and staggered irregularly up and down the hillsides. On my side of the hill, the east side, there was an M60 machinegun bunker covered with rough-cut logs and a couple of layers of sandbags with additional sandbags around the front making it look sort of like the dirt forts we built in sandlots or on construction sites as kids. It was large enough for three or four men and was closer to the top of the hill and higher up than the individual sleeping quarters.

---

[86] *Sometimes spelled "Eagles Nest" or "Eagles' Nest" on various maps and in Army records. It was also sometimes called FSB 1 by some of the units operating in the Dak Pek area.*

Using a borrowed shovel, I improved a previously used hole in a nook that sank back into the hillside under a grassy drop-off and I covered the dirt floor with a thick layer of dry grass. The hole was too small for my sleeping bag, so I opened it up and folded it about in half for a makeshift mattress and pillow. With help from the guy with the shovel, we filled a few sandbags and stacked them to partially close the front of what was becoming a shallow cave. I scrounged up some C-rations and filled my canteen from the water trailer and in about an hour I was an official resident of Eagle's Nest, all moved in and settled. Although I never actually counted the number of men on the hill, today I would estimate around 125, perhaps thirty to forty artillerymen and the balance infantry. I had been on other fire bases but this one was different because it was much smaller and more lightly manned. It was also quieter which was just fine with me.

From my new home, the steep mountainside offered an unobstructed view of the surrounding terrain far below that reached off to the southeast. I was surprised that there was only a single coil of concertina wire located down the hill about twenty meters from my new home but one of the guys said the hill was so steep that the NVA would probably not try a ground attack. Hearing that, I remembered back to the Battle of Dak To just six months earlier when 4th Infantry Division soldiers of the First Brigade (or attached thereto) had successfully attacked and defeated NVA soldiers dug in south of the Dak To forward base on hills named 875 and 1338 by climbing up steep hillsides. Even so, I slept soundly that first night on Eagle's Nest.

The crest of the hill was less than fifty meters wide and not very flat. As I now recall, it was just large enough for only three 105 howitzers, but I am yet unable to find solid documentation regarding the number of artillery pieces on the hill or even exactly which artillery and infantry units manned the firebase at that time.

Charlie battery, 4th Battalion, 42nd artillery was in the Dak Pek area during that time in direct support of part of the 1st Battalion, 22nd Infantry[87] and OPCON to the First Brigade although they were normally part of the Second Brigade.

---

[87] *Lieutenant Colonel Albert B. Akers, Operational Report and Lessons Learned (30 April – 31 June 1968) indicates that C battery, 4/42nd artillery was in the Dak Pek AO under control of the First Brigade from 7 May 1968 until 24 June 1968.*

The view from my new home on the south side of Eagle's Nest. (Steve Stark photo)

The Commo guys had laid a field phone lines from the TOC to various places on the perimeter including a small sandbag bunker on the north end of the hill and just inside the wire where I spent most of my time while on Eagle's Nest with two to three others. Usually beginning before sunrise each morning, we watched the valley below for movement using binoculars. We also occasionally heard chatter on the handset of a PRC-25.[88] For a better view, I often perched on top of the back of the bunker during daylight hours but stayed inside when it was dark.

Looking northeast into the valley from Eagle's Nest (Steve Stark photo)

---

[88] *A field radio commonly used by infantry and artillery in the field.*

For company, we also enjoyed the presence of infantrymen that were positioned on both our flanks and about twenty meters away and a little further down the hill in two small log and sandbag bunkers at the military crest. On some days, we were relieved by other observers between noon and around three but most of the time I remained there all day until after dark. I was given no other duties while there and my job was simple: Observe and report – keep an eye on the jungle and into the valley below and report any changes. Nothing else. Simple and easy work that anybody could do even without any training. My "special assignment" was boring and I wished I had remained at Dak To.

A Sergeant stationed on Eagle's Nest while I was there, a good guy and always happy but whose name is now forgotten. (Steve Stark photo)

From our high perch, the many valleys and hills around Dak Pek were easily visible except when the fog interfered which was usually during the early morning hours. To the northeast at an angle of about forty degrees downward from the horizon (about 700 mils)[89] and five kilometers distant (3 miles), we could see Route 512 which was just a dirt road that ran generally north and south past Dak Pek and paralleled the Dak Poko River. Both the river and the road could have been followed south and then east right into our camp at Dak To with river immediately south of our Recon/Survey bunker. Route 512 was the same little dirt road that ran past the air strip at Dak To.

[89] *In the field artillery, units of angle called mils are used rather than degrees which are not precise enough for artillery fire especially at longer ranges. There are 6,400 mils in a circle, so roughly 17.7777 mils equal one degree.*

Another of the good guys stationed at Eagle's Nest, a Spec 4 whose first name may have been Ray. (Steve Stark photo)

To our north and a little westward at a linear distance[90] of about four and a half kilometers was another fire base called Crow's Nest which was about 150 meters lower than Eagle's Nest. We could sometimes see activity there and listened to their occasional radio transmissions. I'm sure the guys stationed there were watching us and listening to our radio chatter as well.

One afternoon we received a report of a suspected NVA camp or assembly area located at coordinates just to our southwest and only about three kilometers away on the side of a jungle-covered hill. We could easily see the location from the west side of Eagle's Nest and the decision was made to fire two of our 105s into the area and see what happened.[91] After the FDC calculated the initial firing solution, both gun section crews immediately prepared about a half-dozen HE rounds each, and then delivered the initial shots into the hillside followed by other volleys in a dispersion pattern as determined based on adjustments made visually by the Battery Commander. On the third or fourth salvo, a dull secondary explosion erupted, followed immediately by another that produced a bright flash and black smoke. Our gunners had hit a cache of munitions that would hopefully ruin Charlie's day. As the final rounds were fired, everybody on our hill was watching, and the place came alive with whoops and hollers and broad smiles – there are few things on earth much better than American GIs celebrating the success of a job well done.[92] Everybody understood that the ordnance and explosives used by the NVA had to be hauled down the Ho Chi

---

[90] *The line of sight distance, called linear distance, is shorter than ground distance except on flat terrain. The more irregular the terrain, the greater the ground distance.*
[91] *This is called "recon by fire" and basically means that you shoot into the jungle and see what happens. Our infantrymen employed similar tactics using small arms.*
[92] *Although Vietnam field artillerymen in the Central Highlands often employed direct-fire, they seldom saw the results of their work and secondary explosions were always special.*

Minh Trail all the way from the north. There is no telling how many lives were saved that day from what some said was just a lucky shot, while others insisted it was their extraordinary marksmanship. I never learned from whence the intelligence originated for that fire mission, but it must have been either a LRRP team or Special Forces on patrol.

I do not recall any other fire missions being shot from Eagle's Nest during the remainder of my time there which was a total of between two and three weeks.

I usually returned to my cave about two hours after dark and never had to pull any night perimeter guard or other duty and we were never probed or mortared, so I stayed caught up on sleep. At night in the mountains it got cold, but mornings were especially bad and always damp. I had forgotten to bring my field jacket, but the RTO at the TOC requested it for me during a SITREP[93] and somebody at Dak To sent it up on a resupply chopper in a laundry bag with my name on it along with a pair of clean socks and a couple of Hershey bars (with almonds)[94] stuffed in a pocket.

One afternoon near the end of my time on Eagle's Nest, I was given the afternoon off and just lounged around while writing letters home – everything was quiet as usual. About an hour before sundown someone hollered for me to go up to the TOC. I grabbed my gear and scurried up the hill and when I got there the Captain was on the radio talking to someone on the other end whose voice was unsteady and obviously excited, but not quite panicked. The man on the other end spoke in an almost high-pitched whisper as he finally calmed down and gathered his wits to describe what was happening. They had received heavy mortar fire and some small arms fire, and he was convinced a ground attack was imminent. He had no idea the enemy's strength but thought they were coming mainly from the south and probably at least a company-sized element, maybe larger.

A Spec 4 in our TOC was immediately on the radio talking to some unknown firing battery – not ours. The next thing I remember is a Lieutenant in the TOC tapped me on the shoulder and said, "let's go" and he hurried out of the TOC and trotted to the north end of the hill to our observation post with me close behind. When we got there, the Lieutenant cranked the field phone and reported that we were on station. To our north on a hill I could see occasional flashes and it became immediately clear that the NVA attack was from generally the left side of the location – from the west and perhaps the south as had been reported.

---

[93] *At least twice daily, the Commander on Eagle's Nest called in a situation report (SITREP) to Dak To. We could occasionally hear the radio transmissions from Dak To, but messages usually had to be relayed through other locations. All our SITREPS had been about the same while I was there because so little was going on.*

[94] *Hershey bars with almonds were my favorite and my Mom would usually include a few in her care packages. I'm guessing one of my Recon/Survey buddies got them from my stash and put it on the chopper with my field jacket and clean socks.*

In just another few minutes, but what must have seemed an eternity to the guys on the hill under attack, what looked like a fireworks display erupted about a half-click from the top of the hill. The multiple, quick flashes were followed shortly by another series of flashes in about the same place. Another few seconds and there was a third burst, this time a little more eastward and a little further up the hill. The Lieutenant said, "firecracker" and I realized that we were witnessing something heard about but never seen – COFRAM, a new type of artillery shell with multiple small grenade-like bomblets packed together and used against enemy troops in the open – code name "Firecracker."[95]

The flashes from the next three volleys were yet further up the hill but still outside where I imagined the perimeter would be, but it was getting dark and I was only guessing. Then, instantly, there was nothing else to do and the attack was broken by the retreating NVA.
The 6th Battalion, 29th Artillery Commanding Officer, Lieutenant Colonel Albert Akers' Operational Report and Lessons Learned (30 April – 31 June 1968), in Attachment 1-5, describes the following action on 13 June 1968:

B, D/1/22 Infantry at Dak Pek, YB 908 690, received a ground attack by an estimated reinforced Company. The attack came from the SW employing SA and B40 rockets. The perimeter wire was cut, but not penetrated. Artillery was employed using firecracker. It was reported that NVA were observed dragging bodies to SW. (7 counted with estimated 25-30 more). Sweep of area revealed 6 NVA bodies on the wire with numerous blood trails and drag marks.

On the morning I prepared to fly back to Dak To, several Slicks full of very tired airborne troops landed one-by-one on Eagles Nest. They spent a couple of hours mostly laying around and chatting in small groups, also eating. One of them spotted me and called my name, a guy from my hometown, Webster Groves near St. Louis. He had been a football star at our school, so the conversation naturally turned to football and who had won the big Webster-Kirkwood Turkey Day game that previous Thanksgiving, a tradition dating back to the late 1800s.

I think it was a little before noon when I finally boarded a chopper that dropped me off at Dak To a short while later. I reported in at the First Brigade TOC and then at the 6th Battalion, 29th Artillery TOC. It felt good to be home again, see all my buddies and read the mail they had laid on my bunk. My remaining stash of Hershey bars (with almonds) was still safe. That night at Dak To I slept very well, once again back in our bunker and safe on the river.

---

[95] *COFRAM stands for "Controlled Fragmentation Munitions" and according to some records was first used in combat in Vietnam by Charlie Battery, 1st Battalion, 40th Artillery on 12 February 1968, although veterans of that unit I have contacted have yet been unable to confirm their use of it. Another report I read suggests that U. S. Marines first used it someplace near the DMZ.*

While searching for additional information about Dak Pek, I located an article written by Jim Eckl titled *Memories of a 4/42 FA Forward Observer 1968* which says he spent most of his time with Charlie Company, 1st Battalion, 22nd infantry mentioning Dak Pek and Eagle's Nest, also telling of a ground attack on Crow's Nest but does not give a date or talk about using Firecracker. I would very much someday like to visit with Jim if he could ever be located.

During the Vietnam war there were thousands of operations like those on Eagles Nest and around the Dak Pek area in June of 1968, the majority becoming just a line or two in an operational report or maybe scribbled into someone's notebook or more probably just lingering as a faint memory many decades later. Like all wars, most such events are never written about because they were insignificant in comparison to the many much larger and more impressive operations.

For many years after returning home, I would sometimes think about Eagles Nest and search books and magazines when time and resources permitted to look for information about Dak Pek in June of 1968. But I found nothing.

Beginning in the late 1900s during reunions and other Army veteran gatherings, I would sometimes mention Eagle's Nest and Crow's Nest and Ranger's Roost to the guys there, but nobody ever heard of those places. During the 4th Infantry Division national reunion in 2011, I even took photos of Eagle's Nest and the two guys I served with while there, and a couple of hand-drawn maps I had made from memory, but to no avail. The Internet wasn't much help either which was frustrating because it seemed that everything else about the Vietnam War could be found there.

Then, during our visit to NARA as Bob Barrett and I were rapidly turning pages and trying to decide which documents to scan, Bob suddenly turned to me and said, "Look, Steve!" He held in his hand a roughly drawn map marked "Enclosure 7" from one of Colonel Akers' reports as battalion commander of the 6th Battalion, 29th Artillery in 1968. The map showed a hilltop labeled "FSB 1, Eagle's Nest that was positioned just southwest of Dak Pek near the Vietnam-Laos border and in the vicinity of other fire bases labeled Crow's Nest and Ranger's Roost.

Later, I also discovered references to Eagle's Nest in Major Crouse's notebook and a few other references but nothing of any great detail.

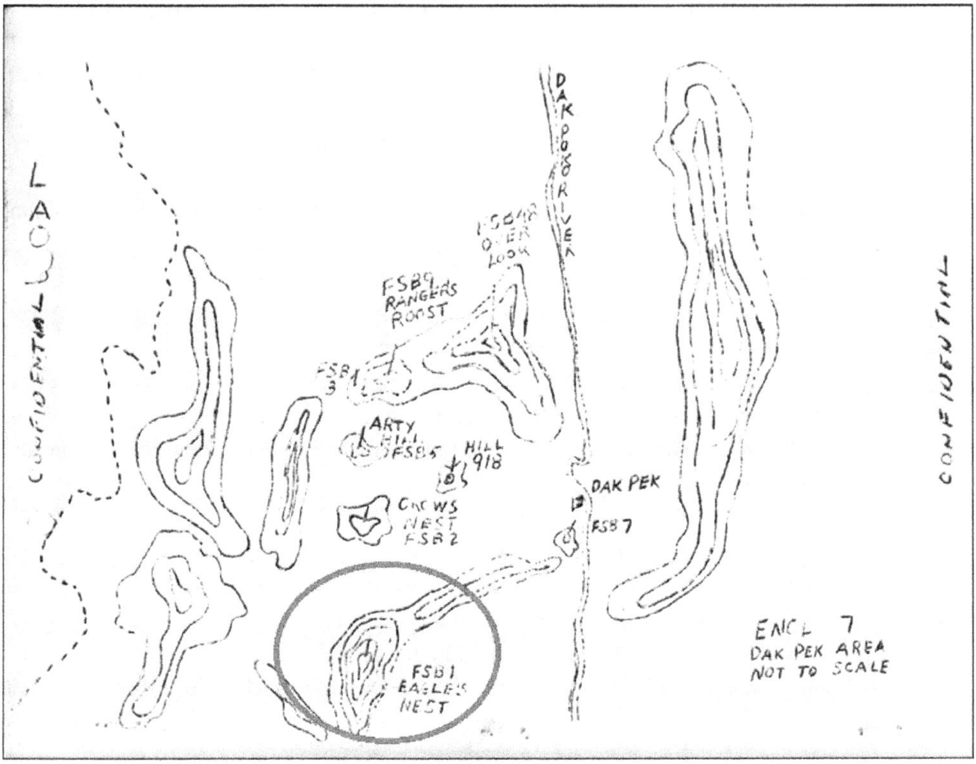

A roughly drawn map of the Dak Pek area included in a senior-level report that shows the general positions of area fire bases including Eagle's Nest (circled in red).

Late June through mid-September 1968 – The Missing Months

After Eagle's Nest and a few other adventures out near the border, I remember very little until the end of my tour of duty. Even so, a few recollections are worth mentioning.

One morning at Dak To while shaving outside our bunker I was approached by a shiny-booted Lieutenant in starched fatigues from MACV in Saigon who said he was in the area and wanted to check on me for my aunt and uncle, Ted and Marjorie Morris, both who worked at that time in Saigon for USAID. I told him I was fine and to tell them "hi" and not to worry.

At some point, possibly late June, we drew lots for our R&R (Rest and Recuperation) and I drew Hawaii while Gary Herder, who was married, drew Bangkok, Thailand. With permission from Sergeant Major Taylor, we switched because Gary's wife, Jean, could much more easily travel to Hawaii than Thailand.

For some reason, around the end of June I stopped keeping my diary and stopped taking photographs. My folks later told me that I did not write home after about the middle of June. I was only twenty-one years old but felt like an old man.

The one event I do remember clearly was on 31 August 1968 when Gary Herder awakened me early one afternoon after I had been up all night. He had terrible news – Major Leslie Crouse had been killed near Ben Het by a mortar round as he and Sergeant Major Charles Taylor were performing a crater analysis. That news sent a shock through our entire unit and brought a dark cloud over me because Major Crouse was a good and decent man and a great soldier who looked out for us and always had a kind word of encouragement even in the worst of times.

I, of course, realized it was getting close to my time to leave Vietnam, and when they offered to let me remain in-country another ninety days for an early Army discharge I turned them down flat. I had seen all I needed to see of Vietnam and the Central Highlands.

When the day finally came, an officer handed me my orders but only gave me about a half hour to get to the chopper pad to fly back to Camp Enari and begin my trip home. Returning to our bunker, there was no time for long handshakes as I said quick goodbyes to all my buddies on the river and immediately walked up the hill past our parapet and BC scope, then across the little dirty road, and then past the First Brigade HQ where I stopped in to say goodbye. At the chopper pad, I boarded a waiting Slick for my last helicopter ride to Camp Enari from Dak To.

Arriving less than an hour later, I caught a ride to Battalion HQ and turned in my rifle and other field gear, then walked down to the connex to retrieve my duffel bag that I'd tossed inside almost a year earlier. My duffel and everything else in the connex made of cloth was coated with a white powdery mildew or otherwise ruined by dark green or black mold. I salvaged my low quarter shoes which were still in pretty good shape and a few other possessions including the brass from a moldy khaki shirt, my M16 expert marksmanship badge and the brass belt buckle I had brought in country which to my amazement was still shiny. The clear plastic bag where I had stored my Missouri driver's license and a little U. S. currency was still safe and sound. Everything else was ruined but no big loss.

After taking a wonderfully warm shower, I dressed in the last clean pair of jungle fatigues I had and put on my boots and soft cap then caught a ride into Pleiku City for a haircut, straight-razor shave, and to get my boots shined – all in the same chair while enjoying a couple of ice-cold beers and chatting with other GIs. I had been to Pleiku City only once or twice before and it hadn't changed much, still noisy and filthy but fairly safe.

The next day I boarded a C-130 (this one with webbed seats) from Pleiku to Cam Ranh Bay and was once again billeted in two-story wooden barracks but in a different part of the base than when we first arrived. They issued me a new khaki uniform and a new garrison cap. The shirt already had Specialist 4 patches sewn on, and I pinned on my M14 Sharpshooter Badge and M16 Expert Badge and the collar insignia salvaged from the connex. They had boxes full of earned ribbons and a new web belt. But there was no 4th Infantry Division patch on the khaki shirt shoulder and instead they handed me a full size green and white 4th Infantry Division patch that was sewn into a clear plastic and white leather holder with a buttonhole for attaching to a uniform shirt button.

I had a hamburger and beer for dinner that night and ice cream for dessert. Then I wrote a letter home and turned in. That night I slept soundly and arose the next morning for a very quick physical exam and an even quicker dental check. In another building, they had us wait in a large room and then interviewed each man separately for about five minutes, asking general questions about our tour of duty while taking notes.

I received two sets of orders, one for a thirty-day leave and the other to report to Fort Sill near the end of October for my last six months of active duty. They had me sign for some travel pay in U. S. dollars then had me fill out more paperwork – always paperwork which was the life blood of the U. S. Army. I took a bus to the Air Force area and boarded a commercial jet for the long ride home. As the plane lifted off, there arose a deafening cheer as everybody smiled and the stewardesses beamed with delight. I was seated near the back of the plane in a middle seat between a giant Marine who had served on the DMZ and a little Navy Seabee who said he spent his tour way down south in the delta. We didn't talk much as we had very little in common.

The stewardesses served us fresh fruit and milk and a couple of good meals on the flight home and we finally landed somewhere in Washington state for a bus ride to Fort Lewis where it was overcast, cold and rainy. They put us in tents with wooden floors and Army metal bunks with no mattresses and dirty sleeping bags.

The following day after more physicals and questioning and paperwork, I was finally given my airplane tickets home and took an Army bus to the civilian airport. Like the other GIs I spotted on the three flights home that day, I was wearing my khakis and low quarters. In the airports and on the flights back to St. Louis nobody ever said a single word to me – not one word – except the clerks who took the tickets and of course the stewardesses.

At Lambert Field in St. Louis, my family met me with hugs and tears. Dad carried my travel bag to the car and Mom would not turn loose of me. We passed a small group of hippies in the terminal, but nothing was said as Dad scowled at them but did not attack. Dad had no use for what he called the "hippie-types" and he disapproved of their very existence.

I found my old bed at home uncomfortable, but still slept very peacefully the first two nights by placing two old bedspreads on the floor. Soon, however, I became acclimated to the soft feel of a real mattress with clean white sheets.

At first the stateside food tasted strange and all my surroundings seemed to smell like soap. For a while it seemed odd drinking from one of Mom's little coffee cups.

Some Americans were especially cruel to returning GIs during the war and even long afterwards. Many Hollywood filmmakers, some historians, and most television and newspaper people still portray us as almost anything but what we were – just normal Americans who answered the call.

In more recent years, it seems that a new kindness toward Vietnam veterans has emerged which some experts have suggested is a form of seeking forgiveness – perhaps an attempted "soothing of the soul." Maybe so. Who knows for certain?

Like most Vietnam veterans, I seldom talk about the war except with people I trust and some of my family and, of course, the guys I served with.

Writing this book has been a challenge and sometimes an uncomfortably difficult journey back in time, but hopefully worth the effort.

Bibliography

Akers, Albert B., *Major General U. S. Army (RET) – Lieutenant Colonel*, Commanding Officer, 6th Battalion, 29th Artillery, First Brigade, 4th Infantry Division (1968). Many personal conversations and correspondence (1999 – 2006).

Alesi, Michael R. (Mike) Reconnaissance and Survey Section, Headquarters Battery, 6th Battalion, 29th Artillery, First Brigade, 4th Infantry Division, Republic of Vietnam, Central Highlands, 1967 – 1968. Many personal conversations and correspondence.

An American Amnesia, How the U. S. Congress Forced the Surrenders of South Vietnam and Cambodia by Bruce Herschensohn, ISBN 978-0-8253-0632-7, published by Beaufort Books, New York, NY.

Barrett, Robert W. (Bob), Lieutenant Colonel (Ret.) – Captain, Artillery, First Brigade, 4th Infantry Division, Central Highlands 1967-1968, serving first as LNO to 5th Battalion, 16th Artillery (September to November 1967), then serving with the 6th Battalion, 29th Artillery first as B Battery Commanding Officer (December 1967 to March 1968), then Battalion S4 from April until May 1968, and then as Headquarters Battery Commanding Officer from June to September 1968. Many personal conversations, discussions, and personal correspondence (2006-2018).

Behind the Burma Road by William R. Peers and Dean Brelis, Copyright 1963, First Published in Great Britain 1964, Robert Hale Limited, 63 Old Brompton Road, London S.W.7, Printed in Great Britain by Clarke, Doble and Brendon LTD. Cattedown, Plymouth.

Campbell, Louie – Cannoneer with Bravo Battery, 6th Battalion, 29th Artillery, First Brigade, 4th Infantry Division, Republic of Vietnam, Central Highlands, 1967 – 1968. Many personal conversations and correspondence.

Crouse, Leslie D., Major (later Lieutenant Colonel), S3, Headquarters Battery, 6th Battalion, 29th Artillery, First Brigade, 4th Infantry Division, Central Highlands 1968 (KIA 31 August 1968) – through his personal memo book and certain parts of letters home (unpublished) as kindly provided by his family and through several personal discussions and correspondence with his sons, Dewayne O. ("Doc") Crouse and John Crouse (2014-2019).

Dak To (Sky Soldiers in South Vietnam's Central Highlands) by Edward F. Murphy, ISBN 0-671-52268-X, First Pocket Books printing (a division of Simon & Schuster Inc., New York, NY) March 1995.

Dieumegard, Theodore J. (Ted) – Reconnaissance and Survey Section, Forward Observer, Aerial Observer, Headquarters Battery, 6th Battalion, 29th Artillery, First Brigade, 4th Infantry Division, Republic of Vietnam, Central Highlands, 1967 – 1968. Many personal conversations and correspondence.

Eckl, Jim, Lieutenant, Forward Observer, 4th Battalion, 42nd Artillery, 4th Infantry Division, Internet article titled *Memories of a 4/42 FA Forward Observer 1968* regarding his experienced in the Dak Pek area.

Farasy, Glen, A., Reconnaissance and Survey Section, Headquarters Battery, 6th Battalion, 29th Artillery, First Brigade, 4th Infantry Division, Republic of Vietnam, Central Highlands, 1967 – 1968. Many personal conversations and correspondence.

Grunt – A Pictorial Report on the US Infantry's Gear and Life During the Vietnam War 1965-1975, Antonio Arques, Andrea Press, ISBN 978-84-96658-49-3 (2014).

Handbook for U. S. Forces in Vietnam, DoD GEN-25/DA Pam 360-621/NAVPERS 15222/AFP 190-1-2/NAVMC 2612, Armed Forces Information and Education Department of Defense (issued 1967)

Harton, Raymond A. (Ray) – Captain, Artillery (LTC Ret.), 6th Battalion, 29th Artillery, 4th Infantry Division, Central Highlands 1966-1967, serving as LNO, 1st Battalion, 8th Infantry. After Army retirement, serving as 6th Battalion, 29th Artillery Search Coordinator 1996 – present. Many personal conversations, discussions, and personal correspondence (2005-2018).

Homyak, Robert (Rob), Reconnaissance and Survey Section, Headquarters Battery, 6th Battalion, 29th Artillery, First Brigade, 4th Infantry Division, Republic of Vietnam, Central Highlands, 1967 – 1968. Many personal conversations and correspondence.

Kvasager, Roger D., Reconnaissance and Survey Section, Headquarters Battery, 6th Battalion, 29th Artillery, First Brigade, 4th Infantry Division, Republic of Vietnam, Central Highlands, 1968-1969. Personal conversations.

Massey, Harold D., Reconnaissance and Survey Section, Headquarters Battery, 6th Battalion, 29th Artillery, First Brigade, 4th Infantry Division, Republic of Vietnam, Central Highlands, 1968 – 1969. Personal conversations and correspondence.

National 4th Infantry Division (Ivy) Division Association – Various information and articles included in "The Ivy Leaves" newsletters as published by the Association and on the various 4th Infantry Division Web sites with their permission.

National Archives and Records Administration (NARA), The National Archives at College Park, Maryland which preserves historical records from the Vietnam War including the following documents: *Quarterly Open Report, 6th Battalion, 29th Artillery, 22 January 1968* by Donald E. Kenney, Lieutenant Colonel, Artillery, Commanding. *Operation MacArthur – Tet 1968, 3rd Battalion, 12th Infantry, 31 January 1968 – 28 February 1968* by Jamie R. Hendrix, Lieutenant Colonel, Infantry. *Operational Report and Lessons Learned, 6th Battalion, 29th Artillery, 30 April 1968 – 31 June 1968*, by Albert B. Akers, Lieutenant Colonel, Artillery, Commanding. *Operational Report, 6th Battalion, 29th Artillery, 31 October 1968* by *William W. Maurer, Lieutenant Colonel, Artillery, Commanding.*

On Guerrilla Warfare, Mao Tse-Tung, Translated from the Chinese and with an introduction by Samuel B. Griffith II, ISBN 978-0-252-06892-8, University of Illinois Press, Champaign, IL.

Pacific Stars and Stripes newspaper, Vol. 24, No. 135, Thursday, 16 May 1968, page 7 story about Recon/Survey Section of 6th Battalion, 29th Artillery building underground bunkers.

Reilly, Patrick J., Reconnaissance and Survey Section, Headquarters Battery, 6th Battalion, 29th Artillery, First Brigade, 4th Infantry Division, Republic of Vietnam, Central Highlands, 1968 – 1970. Many personal conversations and correspondence.

Roden, Allen J., Reconnaissance and Survey Section, Headquarters Battery, 6th Battalion, 29th Artillery, First Brigade, 4th Infantry Division, Republic of Vietnam, Central Highlands, 1967 – 1968. Many personal conversations and correspondence.

Roush, Gary, *of the Vietnam Helicopter Pilots Association* (VHPA) for their permission to use information included on their web site (https://www.vhpa.org) also specifically relating to details regarding pilot CWO Orrin Dyer (KIA 3 February 1968). Also, personal correspondence and telephone communication.

Skoglund, Lawrence L. (Larry), First Lieutenant, Recon/Survey Officer and Forward Observer, Headquarters Battery, 6th Battalion, 29th Artillery, First Brigade, 4th Infantry Division, Central Highlands (1967-'68). His personal recollections (previously unpublished) based on many telephone conversations, personal correspondence and personal discussions.

Stolen Valor – How the Vietnam Generation Was Robbed of its Heroes and its History, by B. G. Burkett and Glenna Whitley, ISBN 1-56530-284-2, Verity Press, Inc., Dallas, TX (1998).

Taylor, Charles A., Sergeant Major, Headquarters Battery, 6th Battalion, 29th Artillery, First Brigade, 4th Infantry Division (1967-'68). Many telephone conversations and personal correspondence (1997 – 2006).

The Battle of Dak-To Steadfast and Loyal, by Robert Quinn, ISBN 078-0-9984115-0-7, SlapDash Publishing, LLC, Hempstead, N.C. (2005). Also, many telephone conversations and personal correspondence and review of certain parts of this book and for his comments and guidance. (2018-2019).

Tyler, Timothy N., Reconnaissance and Survey Section, Headquarters Battery, 6th Battalion, 29th Artillery, First Brigade, 4th Infantry Division, Republic of Vietnam, Central Highlands, 1968 – 1969. Many personal conversations and correspondence especially relating to convoy duty.

Unfit for Command, by John E. O'Neill and Jerome R. Corsi, ISBN 0-89526-017-4, Regency Publishing, Inc. An Eagle Publishing Company, Washington DC (2004).

Unheralded Victory – The Defeat of the Viet Cong and the North Vietnamese Army 1961-1973, by Mark W. Woodruff, ISBN 0-89141-866-0, Ballantine Books, New York, NY and published by The Random House Publishing Group.

VFW Magazine, March 2006, article *Dak To: 33 Days of Violent, Sustained Combat by* Tim Dyhouse (pages 32 to 35).

Vietnam Order of Battle, by Shelby I. Stanton, ISBN 0-88365-709-0, published by Galahad Books, New York, NY.

We Were Soldiers Once…and Young, by Lt. Gen. Harold G. Moore and Joseph L. Galloway, ISBN 0-345-47581-X, published by The Random House Publishing Group, New York, NY.

The following include some of our history and recollections written by Bob Barrett.

ROBERT W. BARRETT

SELECTED MEMORIES OF MY YEAR IN THE VIETNAM CENTRAL HIGHLANDS

HISTORY OF THE CENTRAL HIGHLANDS REDLEGS

FORWARD

It has been a half century since our "Band of Brothers" served together in the Republic of Vietnam (RVN) as soldiers with the 6th Battalion, 29th Artillery, 4th Infantry Division. Our team of Redlegs were part of the second wave of Ivymen after the Division's first year of valiant service, this new "Band" serving roughly from September 1967 until September 1968 with some men arriving and departing earlier or later depending on how the United Stated Army decided to deploy us. The term "Redleg" derives from Army uniforms in the 1800's that were predominantly blue, so to distinguish among the different branches of the Army and to enhance esprit de corps, colored stripes were sewn on the length of the pant legs. The Artillery stripe was red, so artillerymen became known as Redlegs! Our Division was the 4th Infantry Division, one of the oldest in the U.S. Army. It fought in France in World War I. The shoulder patch worn then was the roman numeral for 4, which was IV, or "Eye Vee". From this, the familiar "Ivy" came to be part of the lexicon and lore of the Division, and its Soldiers "Ivymen".

Through letters, telephone calls and the Internet, Lieutenant Colonel Raymond A. Harton, US Army, (Retired), inspired us to begin communicating which resulted in many reunions with our fellow Soldiers. Were it not for Ray Harton, most of us would never have found each other. Ray, by the way, deployed and served with the 6th Battalion, 29th Artillery during its first hard year of combat in Vietnam in 1966-67 including the Valley of Tears during which three Soldiers earned the Medal of Honor and many others earned medals and citations. The reunions led to further inspiration for us to capture our collective experiences in a published history for reading by men who served, and equally by our families and friends.

Our history is in no way to be considered an "official history", although much of the information sources are taken from official 6th Battalion, 29th Artillery and 3rd Battalion, 12th Infantry records on file at the National Archives, Adelphi, MD. The contributing writers served in "B" and Headquarters and Service Batteries, so the history will be from their perspective. Our goal is simply to tell our Vietnam story to share with others.

DEDICATION

I am forever indebted to my fellow Soldier and Co-Author Steve Stark as we persevered together writing our history. We dedicate this book to our Redleg "Band of Brothers" with whom we proudly served in Vietnam. When our Country needed us, we answered the call!
Sadly, on 3 September 2021, Steve Stark made his last formation. Without his encouragement and commitment to honor and tell the story of Artillerymen in the Central Highlands of Vietnam, this book would not have been possible. Thanks Steve!

You will note that his book is a compilation of personal accounts of individuals. In addition to those by Steve and myself, Patrick J. McNulty, Michael S. George, Austin D. Nixon, and John C. Nicolson provided accounts which I included as written. Others provided information in the form of reports, notes, and many conversations which I used to enhance the picture of our collective experience in Vietnam. They are Ray Harton, Albert B. Akers, Donald Kenney, Donald Buchanan, Louie Campbell, Ravin Carpenter, William Davis, Samuel Farris, Nickolas Gillen, Ronald Haas, Larry Manny, Howard McEwen, Charles Snyder, James Williams, Chris Wright, Richard Wright, and John Tempel. I am deeply grateful for all of the input provided by these people. If I erred in using or misrepresented any information provided I take full responsibility for any mistakes incurred.

Finally, I thank my wife Judy who encouraged me through eight years of writing this book. After attending a 6th Battalion, 29th Artillery reunion early in the writing process, she stated after the event "Your Soldiers are what make America Great. You need to tell their story".

THE 4TH INFANTRY DIVISION

The 4th Infantry Division is a storied "workhorse" Division of the US Army. Its Battle Streamers include combat during World War I and World War II in Europe at the time of its deployment from Fort Lewis, Washington in August, 1966. The Division's three Brigades arrived at different locations in the Republic of Vietnam (RVN) by ship, but eventually all Brigades collocated in Pleiku, RVN in the II Corps Area of Operations (AO) in the Central Highland region of Vietnam, establishing a base camp near Dragon Mountain named Camp Enari in honor of First Lieutenant Mark N. Enari, Platoon Leader in "A" Company, 1st Battalion, 12th Infantry. LT Enari was the first 4th ID trooper earning the Silver Star during the Vietnam War. LT Enari was Killed in Action (KIA) as he courageously placed himself in harm's way in order to observe and direct precise artillery fire against enemy positions that had pinned down wounded American Soldiers. The 4th Infantry Division's "Baptism of Fire" in that first year of combat in the largest AO in RVN placed the Division among the ranks of those most honored. During the 4th ID deployment in Vietnam twelve Ivy Soldiers earned the Medal of Honor.

GETTING TO VIETNAM AND THE 4TH INFANTRY DIVISION

Unlike the 4th Infantry Division Soldiers who deployed from Fort Lewis by ship, the second wave of Ivy Soldiers arrived as individual replacements flying in on chartered commercial air from the States. My trip began with arrival at Oakland Army Terminal, California, near San Francisco. The process took a couple of days and included being issued jungle fatigues, the first real proof that I was headed to Vietnam. The flight heading for Vietnam left at night from Travis Air Force Base. It was a TWA Boeing 707, complete with pretty stewardesses. We all wore starched khaki uniforms---no jungle fatigues yet! We made a couple of refueling stops on the way, Hawaii and the Philippines. As we approached Vietnam and began to descend, two things became evident: 1) A tropical environment with lots of vegetation, palm trees, and primitive dwellings, and 2) Bomb craters from bombing raids. I saw no military equipment or soldiers, but I knew I was entering a war zone!

We landed at Ton Son Out Air Base, not far from Saigon. When we departed the airplane, I spotted one of the stewardesses wiping tears from her eyes. She knew it was likely some of us would not be coming home alive. Many retired stewardesses relate that story today, still choking up at the thought. Many men remember them treating us with special care on the way in country, others later celebrating with us when we were headed home. We boarded Army busses with chain-link barriers on the windows, probably to protect against hand grenades and satchel charges, then drove through civilian neighborhoods, arriving at the replacement station in Bien Hoa for in-processing. That took about a day, and where the khakis came off, and the jungle fatigues went on. My assignment to the 4th Infantry Division was confirmed, and soon I was aboard a C-130 aircraft making "milk run" stops at various landing strips, ultimately arriving at Pleiku Air Base. There our group of new (and green) replacements was met by a 4th Infantry Division 2 ½ ton truck which we boarded with duffel bags. I believe we were part of an escorted convoy which departed Pleiku Air Base heading for the 4th Infantry Division Base Camp at Camp Enari, not far away. The short trip over a bumpy dirt road gave me the first impression that Vietnam was different from any place I had experienced before: Vietnamese civilians staring at us, some not looking very friendly, lots of slummy poverty in a tropical environment.

CAMP ENARI, 4th INFANTRY DIVISION BASE CAMP

When we passed through the gate of Camp Enari, I was surprised to see how built-up the compound was, with permanent wooden buildings, all surrounded by 55-gallon drums filled with dirt and supplemented with sandbags, offering some protection from enemy mortars, artillery and rockets. Each battalion had its own area, with headquarters, barracks, mess hall, and supply and motor pool buildings. We were taken to the 4th Infantry Division Replacement Battalion, near the center of Camp Enari, for in-processing into the Division. This took about half a day, and then each of us was transported to our respective units, "bag and baggage". (Editor's Note: Captain Barrett initially assigned to 5th Battalion, 16th Artillery, the General Support Battalion in the 4th Infantry Division Artillery equipped with the M109 155 mm Self Propelled Howitzer, and the M110 8-inch Self Propelled Howitzer.) None of us new arrivals were acclimated to Vietnam time, terrain, or climate. It was all new to us. To help "break in" new replacements, the 4th Infantry Division Leadership established a five day "in country" orientation program to ease the acclimation process. This included zeroing of our personal weapons – the relatively new M16 rifle. The highlight of this training was the conduct of "sweep" patrols outside of the wire fence surrounding Camp Enari. All of the training, including the patrolling, was overseen by experienced Division cadre. The orientation course was a good way to start off a year in Vietnam, helping build confidence.

In base camp two unique features quickly became evident: the latrines, and the personal wash basins. Let's begin with the latrines which were "out houses" but were known to all of us as "the shitters". These were separate "out" buildings, with the interior set up with various numbers of "holes" for one to sit on. Underneath each hole was half a 55-gallon drum to catch the human waste. In the more civilized areas, like our base camps, most latrines were constructed with hinged trap doors underneath so the filled barrels could be hooked using a length of bent pipe, wire, or piece of bent iron rod (sort of like a long hay hook) and pulled out. In base camp this was done by local hire Vietnamese civilians, affectionately known as "shit burners" who would pull the full barrels out, pour in some kerosene, and burn it off while occasionally stirring. Someone had to do it, so in the more remote locations our enlisted Soldiers were responsible for that filthy task and everybody took his turn. By the way, one of the pleasures of visiting the shitter was to catch up on the latest news in the Stars and Stripes newspapers thoughtfully left behind by those who preceded you. The shitter took care of the number "2" needs, and for number "1" we had another unique device called the "piss tube". These were beautifully simple, either empty 105mm, 155mm, or 8-inch hollow howitzer powder canister tubes placed at a convenient angle about halfway into the ground, beneath which was gravel (when available), and sometimes sand or loosened dirt. They were sometimes supported by concrete tops and could be sanitized by occasional burning using a mixture of diesel and gasoline. Powder lye was sometimes poured down in lieu of burning to help keep the stench down and probably to disinfect as well. The tubes were occasionally abandoned and even moved to prevent undue contamination in areas occupied for longer periods of time. Sometimes, a wire screen was placed over the tube, but the reason is still a mystery. Field artillery units had plenty of these canisters so piss tubes were readily available in all of our future locations. We were really spoiled. There was one downside to the piss tubes---bees seemed to love the urine, so they were always buzzing around the tubes, causing many a second thought about "hanging it out!"

The portable wash basins were pretty common in Camp Enari. They were brass medal bowls, about 18 inches across and six inches deep. They were very handy for shaving. Most Soldiers assigned for duty at Camp Enari owned one, readily available for purchase "on the economy" from "Mama-San". Having a wash basin as part of your gear meant you didn't have to use your steel pot for the same purpose. Our Soldiers who didn't have the luxury of living at Camp Enari still had to rely on their handy steel pot!

There were a couple of other distinctive identifiers of Camp Enari. Although Pleiku is in the Central Highlands, the immediate area surrounding Camp Enari is relatively flat. One exception was a prominent hill named Dragon Mountain. Because of its "high ground" status, Dragon Mountain was home to many radio antennae. This elevated position facilitated communications in the 4th Infantry Division Area of Operations (AO) by increasing the line of sight to distant radio receivers in this expansive AO.

The perimeter of Camp Enari was ringed with a large barbed wire fence. At approximately every 100 meters there were 50-foot tall guard towers, accessed by stairs, with a covered roof over a standing rectangular platform. From these towers Soldiers performed around the clock watch duty as part of the Base Camp Security Plan. Some of the bunkers were 2-man ground-level placements, about the same distance apart as I recall. The towers mixed with these smaller ground-level structures with interlocking fields of fire to provide a shield of final defensive fires in case of enemy ground attack. I would later learn that there was a provisional firing battery located in Camp Enari to provide artillery support to defense of the Camp. I know that each of the 6/29th Bn firing batteries (A, B, and C) contributed one 105 mm howitzer and crew to the base camp battery.

4TH INFANTRY DIVISION AREA OF OPERATIONS

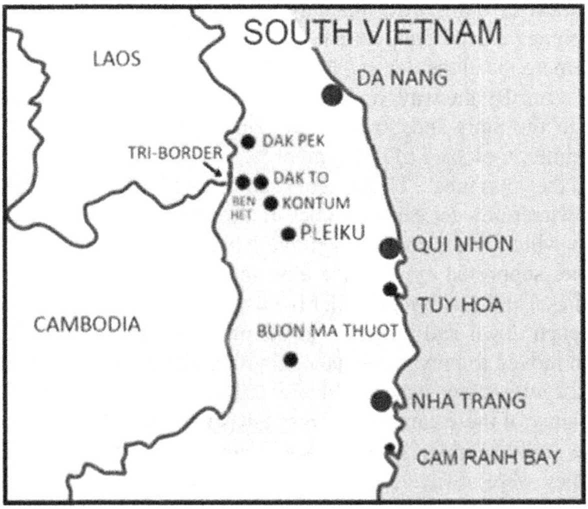

Approximate location of the Central Highlands highlighted in green was largely the responsibility of the First Brigade, 4th Infantry Division in 1967 and '68, mainly the area north of Buon Ma Thuot. Map produced and provided by Steve Stark.

LIFE IN THE CENTRAL HIGHLANDS OF VIETNAM

View of mountains in Dak To area. 1LT Bill Davis photo

The Central Highlands of Vietnam had to be the most desirable region in the country to be assigned, given that you had to fight a war there. The verdant mountains, with triple canopy jungle, some a mile high, had the beauty reminiscent of the Appalachians in the eastern US, complete with clear running rivers and streams. One of the first things a new arrival would notice was a unique Vietnam "smell", probably from the wood fires burning teak and mahogany for cooking. The smell was everywhere, and it was pleasant. Living in the highlands, we enjoyed cooler temperatures than our fellow Soldiers fighting in the lowlands. In fact, it got down right cool at night. A new item of equipment for the Vietnam era, the poncho liner, became very popular with Soldiers. It was quilted, with the outer material camouflage color. They were very comfortable to wrap up in and went a long way to stave off the chill of the night.

There were two seasons in the Central Highlands: Rainy, and dry. The rainy season, known as the monsoons, generally matched our spring season in the US, about April through July when it tapered off. Because we were in the Highlands where drainage was better, we avoided the flooding that impacted some of the lowland areas.

A unique aspect of the Highlands was the dense "triple canopy jungle", a phenomenon not known in the US. Triple canopy describes a mountainous jungle cover of three layers of dense foliage, at approximately 50, 100, and 150 feet above the floor of the jungle. I'm no horticulture expert, so I don't know how the three layers were formed, but they presented a challenge to be reckoned with fighting there. First, they created much darkness on the jungle floor where most of the fighting in the Central Highlands took place. The three layers of canopy blocked out most of the sunlight. Second, the three layers of canopy negatively impacted the effects of artillery fire, particularly that of our 105 mm howitzers. The triple canopy acted as a "triple shield" against the shells intended to impact on the ground, or 20 meters above the ground in the case of a time fuse. Simply stated, the thick jungle canopy caused the point detonating fuses of the artillery shells to pre-detonate at the 150-foot level, or successively at the 100 or 50-foot level, greatly dissipating the killing power of the artillery shells, if not completely negating it. The time fuses, both mechanical, and variable time, were also negatively affected by the jungle. Time fuses are designed to detonate at a 20-meter height of burst, and their primary targets were troops in the open, or unprotected troops in trenches. The mechanical time fuse was set by the cannoneers just prior to firing, calculated by the Fire Direction Center (FDC). The Variable Time Fuse, or VT fuse, was developed during WWII. It employed a mini "radar-like" radio which was pre-set to detonate at a 20-meter height of burst. The three canopy layers would successively further disperse the shell fragments once detonated, and the VT fuse could be "fooled" by the 150-foot canopy, making the fuse "think" it was 20 meters above the ground, rather than 20 meters above the canopy. With either time fuse, the net effect was widely dispersed shell fragments triggered by the premature canopy bursts. If any fragments penetrated through the dense canopy, their effects on the enemy were often minimal.

Triple canopy jungle at a Fire Support Base in the Dak To area. 1LT Bill Davis Photo

Earlier I had talked about the "smell" of the Highlands caused by the wood burning cooking fires. There were other smells as well, but these were blended by the Americans. Our 105-mm battalion fired a lot of artillery rounds from their fire bases, and the exploding propellant in the shell casings gave off a pungent odor only a Redleg surrounded by howitzers on a fire base could appreciate. That smell of cordite worked two ways: Our shells leaving the fire base, and from enemy shells coming in! Another classic smell was that of JP-4. Jet Propulsion Fuel #4 was the fuel that propelled the jet turbine engines of the OH-1 Cayuse, UH-1 Huey, AH-1 Cobra, CH-47 Chinook, and the CH-54 Flying Crane helicopters. Whenever in close proximity to one of these choppers, whether it be on a landing pad at a base camp, pick-up zone (PZ) or landing zone (LZ) at a fire base or in the "boonies" on a combat assault, the smell of JP-4 exhaust was there. It was a distinct and unique smell, not unpleasant at all, and only those who experienced it first-hand can fully appreciate it.

THE HELICOPTER WAR

Bell UH-1 Utility Helicopter (Huey), 1LT Bill Davis photo

Hughes OH-1 Cayuse Helicopter. 1LT Bill Davis photo

Bell OH-23 Observation Helicopter. Soldier name unknown. SGT Charlie Snyder photo

CH-47 Chinook Helicopter with sling load at Dak To destined for a nearby fire support base. SGT Charlie Snyder photo

Fairchild C-123 Caribou transport plane. Source of photo unknown

CH-54 Flying Crane Helicopter delivering towed 155 Howitzers northeast of Dak To. SP4 John Nicolson photo

Vietnam was the first "helicopter war", and especially true in the Central Highlands where steep mountains covered with dense jungle and poor, sparse roads greatly hindered traditional ground movement. The "Airmobile Concept" was a new concept of warfare, built around the use of helicopters to increase mobility of soldiers and equipment in mountainous, jungle terrain such as we fought in the highlands of Vietnam. The rapid technological advancement after WWII led to the development of a new family of helicopters with a spectrum of capabilities ranging from observation, troop and equipment transport, medical evacuation, to attack. The workhorse of the family of helicopters was the Bell UH-1 (Utility) helicopter fondly named the "Huey" by all who used them. The Huey had three key roles: 1) As a transport helicopter ferrying Infantry Soldiers to and from landing zones for combat assaults and extractions; 2) As medevac helicopters, nick - named "Dust Off"; and 3) As gunships armed with mounted M-60 machineguns and 2.5-inch rockets to provide close-in fire support to Soldiers on the ground. The gunships often accompanied the troop carrying Hueys (known as "slicks") to provide fire support on combat assaults, or as overhead security flying along ground convoy routes to provide quick reaction firepower in case the convoy was ambushed along the way.

Thousands of Huey's were deployed in Vietnam. They were often easy targets for the VC and NVA because they were exposed to enemy fire during combat assaults and extractions. Many were shot down. We could not have engaged the enemy effectively in Vietnam without the Huey. The Huey gave us the means to take the fight to him.

The sister ship to the Huey was the Boeing CH-47 Chinook helicopter. This was the large twin-engine cargo helicopter absolutely essential to the creation of firebases where our 105 mm howitzers were employed. The Chinooks lifted our howitzers using slings up to the mountain top firebases, often at very high altitudes. The Chinooks were affectionately known as "Hooks". They also lifted critical ammunition using nets and slings to accompany the howitzer. During moves from one firebase to the next, Chinooks brought in sling loads of essential equipment like the FDC CONEX, the water trailer (aka "water buffalo"), perforated steel planking (PSP) and sandbags for building protective bunkers, and the Soldiers in the hold. When a firing battery departed one fire base to occupy another one, a great deal of teamwork was required among the pilot skillfully hovering the helicopter, the crew chief in the Chinook controlling the hook on the hovering helicopter, and the Soldiers on the ground preparing the loads with slings and nets. The "hook up" man on the ground had to ensure the sling was properly secured to the load, then he had to secure the lift "donut" attached to the sling to the hook dropping down from the helicopter controlled by the crew chief. All of this had to be done while the helicopter was hovering over the hook up man, churning up huge clouds of dust making it very difficult to see. This was a very dangerous, dirty, and noisy job. PFC Louie "Pee Wee" Campbell, B Battery Cannoneer, related his personal experience as a Chinook "Hook Up" man. "We were moving one time and I was standing on top of the FDC CONEX. The Chinook was supposed to pick up the CONEX, and I was going to hook it to the bottom of the Chinook with a round ring (donut). The pilot can't see me so there is a crew chief looking at me through an opening in the bottom of the helicopter. The crew chief is telling the pilot what to do (what a messed-up job!). Well, as you can tell, that pilot moved the Chinook down, and down, and down, and I went on my back lying between the Chinook and the FDC CONEX. The only thing that saved me was one of the chopper wheels sat on top of the CONEX, keeping it from crushing me. Good thing I was skinny. The Chinook lifted off with the CONEX and I just slid off to the ground about ten feet down. Damn, it hurt!"

Chinook Helicopter lifting CONEX for FDC in Fire Base near Dak To, 1LT Bill Davis photo

PFC Louie Campbell relaxing with his howitzer "Brute" on Hill 1338 overlooking Dak To. SGT Nick Gillen photo

I have already described the role of the 'gunships', the Bell UH-1 Huey configured as a gunship, and later the Bell AH-1 Cobra gunship. The Huey gunship was a "cut down" version of the troop carrier, used only in the attack helicopter role. About half way through our year in 'Nam a brand-new helicopter appeared in the skies, the Bell AH-1 Cobra. It was the sleekest, sexiest chopper in the sky! It looked like a shark, and many ground crews painted shark's teeth on the front cowling, a throwback tribute to the WWII P-40 Warhawk of "Flying Tigers" fame. When I first saw a Cobra fly overhead in the Dak To area, it happened to be flying directly toward me. It was strikingly thin in width, looking to be not much wider than the shoulder width of the pilot. The Cobra was armed with machine guns and rockets and was used exclusively in the attack helicopter role.

One of the more novel helicopters in Vietnam was the Hughes OH-1 Cayuse Observation Helicopter. I have no idea what nick name it was given, if any, but to me it looked and maneuvered like a dragon fly. As its OH designation indicates, the Cayuse was the consummate observation chopper. It was a small chopper, easily identified by a distinct round bubble in the front where the pilot sat, and a short off-set tail boom. I observed a couple of them on a mission in the Dak To area south of Hill 1338. I was impressed with their speed and maneuverability. They moved around much like dragon flies would, darting quickly to a spot, then stopping, presumably to conduct observation, then moving quickly to another observation spot, and so on. I also noted that they were very quiet as helicopters go. I attribute this to a high RPM rotor blade that cut down on the classic helicopter "whop, whop, whop" noise.

Another Observation Helicopter was the OH-23. It was light, recognizable by its clear plexiglass "bubble" in front for the pilot and a passenger, with a slender tail "boom" extending aft from the "bubble" out to the tail rotor. LTC Akers, our Battalion Commander beginning in February 1968, has high praise for the H-23 in his "Quarterly Operational Report of Lessons Learned, ending 30 April 1968." "The usefulness of the OH-23 helicopter cannot be overemphasized. The OH-23 is used extensively for artillery adjustment, visual reconnaissance, transportation of material and supplies (to meet our operational requirement) and personnel. In its artillery adjustment role, the OH-23 bridges the visibility gap that exists due to terrain, also providing an Air Observer who is able to react to any situation brought about by intelligence or contact. In the visual reconnaissance role, the OH-23 has on numerous occasions been able to uncover valuable area intelligence, exploit reported activity as well as provide commanders with a reconnaissance means for future locations; however, it is not flown at low level altitudes without gunship coverage due to its vulnerability."

I'll briefly mention one of the more unique helicopters of the Vietnam War: the Sikorsky CH-54 Flying Crane. It was the largest helicopter by far, its primary mission to lift heavy pieces of equipment too large for any other helicopters. In the 6/29th Artillery we never had a need for its services. But it was a regular for the 155-mm towed howitzer (lovingly known as the "Pig"). The 1st/92 Artillery, a 155-mm towed battalion, provided reinforcing fires to the 4th Infantry Division Artillery during the Battle of Dak To. The Pigs were too heavy to be hooked into a fire base by a Chinook, so the CH-54 was tasked to do the job. The Flying Crane had the look of a grass hopper because it had four long, solid legs that straddled the lift area where a formidable crane hoist would be dropped down to pick up the Pig, all four legs straddling it. Once hooked up, the powerful engine could lift the howitzer into the air and deliver it to a fire base to prepare for action.

I would be remiss if I didn't mention some other aircraft that provided valuable and sometimes spectacular support to us in the "Can Do" Battalion. The king of spectacular was "Puff the Magic Dragon", or "Spooky". What a fantastic and clever weapons system this was, a real tribute to U.S. Air Force ingenuity. "Puff" was a WWII era Air Force C-47 "Sky Train" cargo aircraft modified from the workhorse cargo aircraft which famously transported the 82nd and 101st Airborne Divisions to Normandy for the Invasion of Europe. But "Puff" now had a new mission, all designed to provide close fire support to the Soldier on the ground. It was configured with three 7.62 Gatling Miniguns. For night missions "Puff" carried flares for illumination. I'm not sure of the technology used to aim these "Buck Rogers" style weapons, but I can guarantee the pilot and crew had great skill and loved using these gadgets against the enemy. Above all, it was fun to watch "Puff" at work, especially at night. "Puff" would generally circle his designated target area, dipping a wing as part of the aiming process. Then a burst from the Gatling Miniguns would be fired, and you could follow a stream of tracers to the target. Stream after stream would follow as "Puff" circled the target. The bursts looked like a stream of fire from a flame thrower in the night darkness, or more graphically, like someone taking a piss from the plane! Most impressive, and I'm sure feared by the enemy!

If Soldiers on the ground needed emergency illumination at night, "Spooky" could pop off a large volume of bright illumination flares in a short period of time. Those flares could light up a dark night, making the decisive difference when firebases were threatened with being overrun.
The U.S. Air Force also provided the classic form of close air support. It came in two forms: High tech, and Low tech. The high tech came in the form of "fast moving" fighters, usually F-4 Phantoms, but also F-100 Thunder Chiefs. The low tech was provided by Korean War vintage A-1 Sky Raiders, nicknamed "Ho-Bo's". These close support aircraft were guided to their targets by U.S. Air Force Forward Air Controllers (FACs), also skilled fighter pilots themselves. The FAC's flew an OV-1, identified by its two engines, one in front pulling, the one in the rear pushing, and by its twin boom tail. Think of the FAC's as the Air Force equivalent of our Artillery Forward Observers. Our hats are off to our "Brothers in Blue" for the essential support they provided to us on the ground. It may not have always been visible, but it was always there!
Our Air Force provided another kind of support, but not "close air" support. This support came from afar, but we heard and saw it in the "Arch Light" bombing conducted by huge B-52 Bombers.

"Arc Lights" earned their name because these were massive bomb strikes using 500, 750 and 1000-pound bombs dropped from miles high, almost always aimed at targets along the Ho Chi Minh Trail west of Dak To. These bomb drops were usually conducted at night and were so powerful that when the bombs detonated a series of arches of light would flash up in sequence from the target area followed shortly by a noticeable shaking of the ground. And finally, if conditions were favorable and the target was not too far away, you could hear the rumble of the bombs detonating within a minute. I was always glad it was our enemy, and not us, on the receiving end of those Arch Lights. Can't imagine what it was like!

I would be remiss in not mentioning one other Army transport plane used in Vietnam because many of you likely rode in one. It was the C-123 Caribou built by Fairchild, and an Army, not an Air Force plane. Think of it as a small version of the Air Force C-130. It was designed for short take-offs and landings on small, remote airfields, ideal for Vietnam. They all reverted to Air Force control after the Vietnam War.

PRELUDE TO THE BATTLE OF DAK TO, AND TET OFFENSIVE

September and October 1967 were a time of light combat activity for the Division. The Division AO stretched from An Khe to the east, Duc Co (The Oasis) to the west, and Ban Me Thout to the south. (Refer to Map). There were occasional sharp exchanges with the enemy, mostly Viet Cong. The enemy was laying low, the new arrivals in the Division were wondering "where is all the action we had been told to expect?" This relatively quiet time was a blessing for the Battalion. It gave us, most of whom were "newbies" a good chance to build teamwork in our sections, learn our jobs in a totally new environment, and just settle in. These benefits far outweighed the bouts of homesickness and boredom we were experiencing.

But things were soon to change for the new wave of Soldiers in the 4th ID. Soon they would be embroiled in two major combat actions which would define our year in Vietnam. At our "B" Battery/ Headquarters and Service Battery (HSB) Reunion in October 2016 in Nashville I discussed our plans for writing our history with Captain (Later Lieutenant Colonel) Richard Wright who followed me in command of "B" Battery in April 1968. Dick recommended two excellent publications which would help me with my research. They are "Dak To and the Border Battles of Vietnam, 1967-1968" by Michael A. Eggleston, and "The TET OFFENSIVE, A Concise History" by James H. Willbanks. These books enlightened me about the strategic planning being conducted by the highest level of Communist leadership in North Vietnam. By mid-1967 North Vietnam's leaders were concerned that the war in the South was not going well for them. The increasing buildup of US Forces was causing increased casualties for the Viet Cong (VC) and North Vietnamese Army (NVA) forces in the South. Morale was suffering and desertions were on the rise. In April,1967 the Communist leadership in North Vietnam passed a resolution which called for a "spontaneous uprising in order to win a decisive victory in the shortest possible time." It was decided the offensive would be launched in early 1968 during Vietnam's Tet holiday. General Vo Nguyen Giap, North Vietnam's top military leader was given the responsibility to plan and lead this offensive. Giap used these four assumptions for his plan:

- The Army of (South) Vietnam (ARVN) would collapse and desert in large numbers.
- The people would rally to support the VC.
- The ARVN and people would turn on the Americans.
- The tactical situation at Khe Sanh would parallel that at Dien Bien Phu against the French in 1954.

During the fall of 1967 Giap planned a series of "Border Battles" close to the Laotian and Cambodian borders to draw the US and ARVN forces away from population centers such as Saigon, Pleiku, and Hue. This would leave the cities and towns uncovered. The final "Border Battle" occurred near the US Special Forces Camp at Ben Het near an air strip at Dak To northwest of Kontum in the mountainous area along the Laotian and Cambodian borders in the Central Highlands. In late October intelligence indicated that four NVA main force regiments were moving into the Dak To region. General Westmoreland quickly sent US units to Dak To, including the 1st Brigade, 4th ID. The stage was set for the Battle of Dak To. After my research, I now knew that the Battle of Dak To, and the subsequent battles around Dak To during the Tet Offensive were all part of the "Grand Plan" by the NVA.

B BATTERY AT THE BATTLE OF DAK TO, NOVEMBER 1967

PFC Louie Campbell (L) and SP4 Jimmy Williams (R) loaded on "Deuce and a Half" truck departing "The Oasis" and heading north to Dak To. Soldier in middle unknown. PFC Louie Campbell photo.

6th Battalion, 29th Artillery Convoy heading north from The Oasis to Dak To on Highway 14. PFC Louie Campbell photo.

Dak To air strip on final approach by helicopter. SGT Charlie Snyder photo

After a relatively quiet two months in September and October, 1967, in the Oasis area west of Pleiku, the 6/29th Artillery was ordered the end of October to road march north to Dak To. The road march put the battalion on Highway 14, a major North-South road which passed through Kontum on the way to Dak To to join the 1st Bde, 4th ID to counter the NVA threat building there. The battalion closed on Dak To 29 October 1967. Lieutenant Colonel Donald E. Kenney was commanding the battalion at the time.

Key Field Artillery Leadership, just prior to Battle of Dak To: L-R, Captain Jackson, Commander A Battery, 5th Battalion, 16th Artillery, LTC Donald Kenney, Commander 6th Battalion, 29th Artillery, MAJ Olson, Operations Officer, 5th Battalion, 16th Artillery, LTC Henry Evans, Commander 5th Battalion, 16th Artillery. Official US Army Photo courtesy LTC Kenney

He reports in his Operational Report for Quarterly Reporting Period Ending 31 January 1968 that the enemy situation as of 1 Nov 67 was as follows:

24th NVA Regiment, Strength--1285
32nd NVA Regiment, Strength—1600
66th NVA Regiment, Strength—1560
174th NVA Regiment, Strength—1760
304th LF Bn (VC), Strength—520
40th Arty Regiment, Strength—Unknown

In order to better understand the flow of combat actions of B Battery and Headquarters and Service Battery during the Battle of Dak To, and the subsequent Tet Offensive, I am including here three very helpful map diagrams which lay out the location of the Fire Support Bases and key terrain features used by the US Forces, and the NVA and VC during this period of intense combat. These diagrams were included in the Operational Report – Lessons Learned for the Period Ending 31 July 1968 compiled by Lieutenant Colonel Albert B. Akers, Commanding Officer of the 6th/29th Artillery. LTC Akers had assumed command of the battalion from LTC Kenney on 18 February,

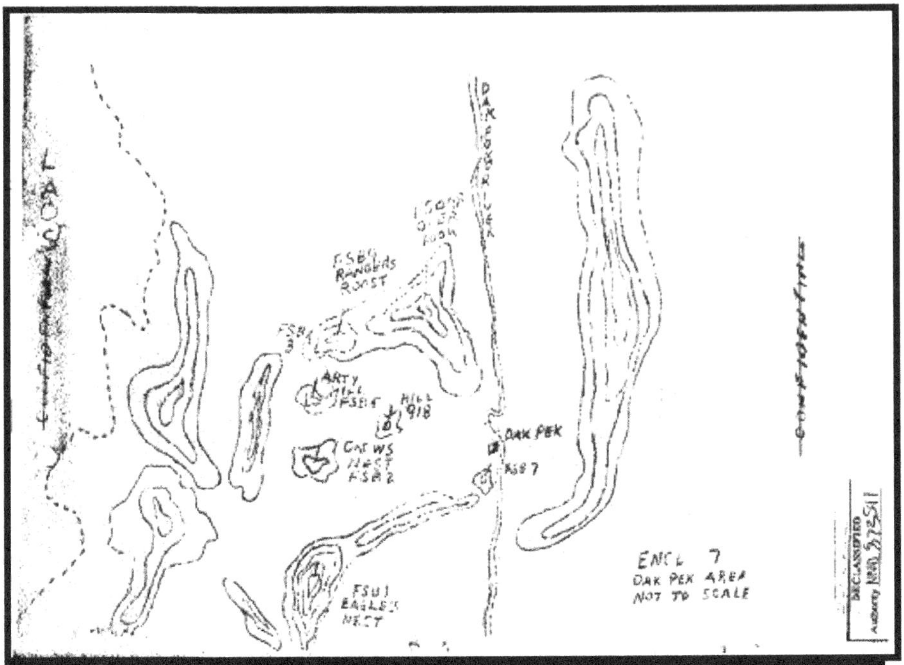
LAOS
FSB 6
RANGERS ROOST
HILL 918
DAK PEK
FSB 7
FSB 1
EAGLES NEST
ENCL 7
DAK PEK AREA
NOT TO SCALE
DECLASSIFIED

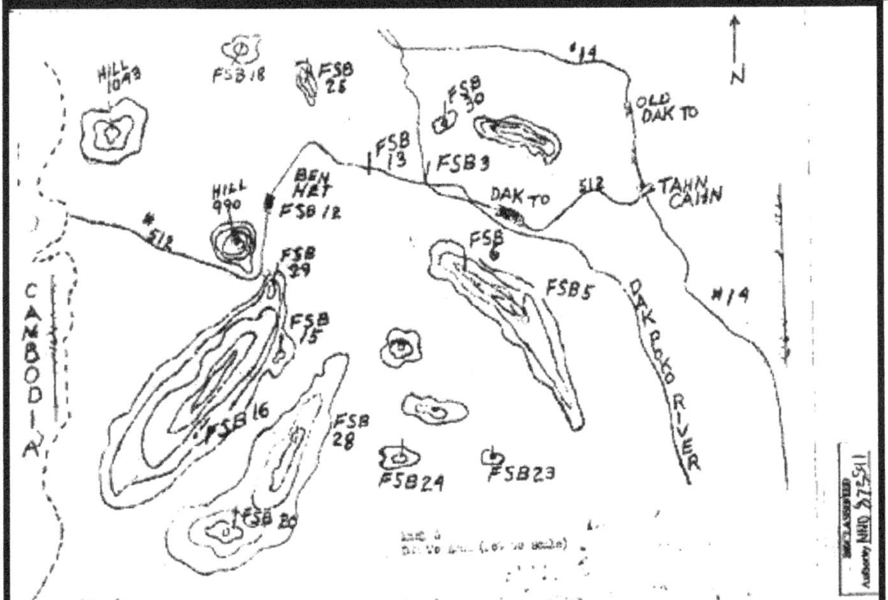
HILL 1049
FSB 18
FSB 25
FSB 30
#14
OLD DAK TO
N
FSB 13
FSB 3
BEN HET
FSB 12
HILL 990
#512
DAK TO
512
TAHN CAHN
FSB 29
FSB
FSB 5
CAMBODIA
FSB 15
DAK POKO RIVER
#14
FSB 16
FSB 28
FSB 24
FSB 23
FSB 20
DECLASSIFIED

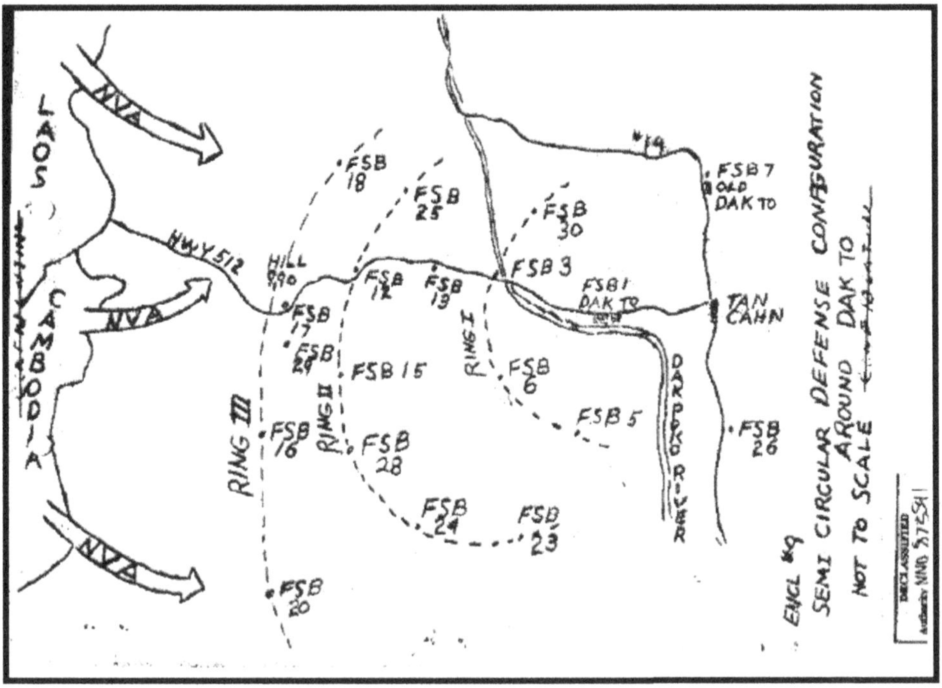

Three sketched maps of terrain features and Fire Support Bases in the Dak To Area of Operations. Descriptions from top to bottom: Top: Fire Support Bases and Observation Posts northwest of Dak To next to border with Laos. Dak Pek is a Special Forces Camp. Middle: Numbered Fire Support Bases surrounding Dak To were key battle positions for US Army units during the Battle of Dak To and the 1968 TET Offensive. Western most Fire Support Bases were within range of the Cambodian Border. Ben Het was a key Special Forces Camp. Bottom: Concentric Defensive Rings with associated Fire Support Bases designed to disrupt and destroy NVA forces attacking from support bases in Cambodia and Laos, intent on overrunning the US Army at Dak To, moving to the coast, and splitting South Vietnam in half. Diagrams from (6th Battalion 29th Artillery) Operational Report—Lessons Learned for Period Ending 31 July 1968.

By mid-November contacts with the enemy were becoming more frequent and deadly. Captain Michael S. George was in command of B Battery and he wrote an excellent article for the Jan/Feb 2009 issue of the Military Order of the World Wars Officer Review magazine. It is entitled "Thirty Days at Dak To—A Field Artillery Story". There is no better description of what this fierce battle was like for the Soldiers of B Battery. Mike and the Military Order of World Wars Officer Review magazine have given their permission to include it in our history in its entirety.

SSG "Buck" Buchanan standing on howitzer trail firing high angle support for attack on Hill 1338, SP4 Sam Farris photo.

PFC Chris Wright, B Battery "Mess" Sergeant (L), making final preparations for Thanksgiving Dinner on Hill 1338, SP4 Sam Farris photo

B Battery cannoneers enjoying Thanksgiving Dinner on Hill 1338. SP4 Sam Farris without shirt. PFC Louie Campbell second from right. SP4 Robert Butler far right. Other Soldiers unknown. SP4 Sam Farris photo.

Captured enemy weapons on Hill 1338. SP4 Sam Farris photo.

Initial occupation of B Battery on Hill 530, 25 Nov 67, SSG Elton Jones (L), Chief of Section, CPL Donald Fink, Gunner, Official U.S. Army photo.

Fire Support Base on Hill 530, west of Dak To, SP4 Sam Farris photo

Mike writes: In November 1967, the 1st Brigade of the 4th Infantry Division moved to Dak To in Vietnam's Central Highlands. Battery B, 6th Battalion, 29th Field Artillery (M101A1, 105mm, Towed), under my command, was in Direct Support of the Division's 3rd Battalion, 12th Infantry

(3-12 IN). It was common at this time for a single battery to provide dedicated support to a battalion rather than operate within a battalion formation, the traditional method of employment. At the time of our deployment to Dak To, intelligence reports indicated that four North Vietnamese Army (NVA) Regiments occupied positions in the hills surrounding the area. Initially, my battery was on the north side of a berm bordering the Dak To landing strip. The 3-12 IN was to assault the NVA occupying Hill 1338, scene of a disastrous Special Forces incursion in July (Allan Clark, *Wounded Soldier, Healing Warrior*). Fighting was intense against an entrenched force. The battery provided continuous fire support well into the night, so much so that our basic load was expended and we received emergency night resupply directly from a 5-ton tractor-trailer. This created a potential problem, as rounds had to be uncrated and unpacked in the dark thus posing a safety and fire hazard due to the accumulating debris while firing continued unabated. Fortunately, no such situation developed. Several thousand rounds were fired during this engagement.

A few days later CH-47 Helicopters airlifted the battery to a ridge line overlooking the Dak To landing strip. There was evidence of enemy occupation in the form of trenches on the military crest and the lingering smell of tear gas. Fighting on the now adjacent Hill 1338 was continuing and the battery suffered its first casualty, one of my reconnaissance sergeants (Forward Observer Party) was killed while directing fire on the enemy. We would remain in this position through Thanksgiving Day before moving south to Hill 530. This would prove to be a most fateful move. It was the sixth move the Battery would make since I had assumed command in September.

Hill 530 was some 15 kilometers southwest of Dak To and was the only prominent terrain feature in a broad forested valley that led to the tri-border are of Vietnam, Laos, and Cambodia to the west. For the first time since assuming command, I had the opportunity to reconnoiter a proposed firing position. I reported to my Battalion Commander that I could not fit the battery on this position due to its small (and narrow) size and the fact that we would be co-locating with the 3-12 IN Headquarters, to include the battalion mortar section. Nevertheless, we were shortly on our way south (our Battalion motto was "Can Do"). An airstrike had felled most of the trees and we used chainsaws to cut and clear the position. On the east end of the hill was a bomb crater some 15 feet deep into which my Fire Direction Center, a CONEX container, was lowered. The guns (Howitzers) stretched out to the west. The crest of Hill 530 was at most 30 meters wide so there was great difficulty in achieving some semblance of a traditional "W" layout, much less the more common "Star" formation designed to facilitate a 6400 Mil (360 degrees) firing capability. No howitzer was further than 15 meters apart. Our personnel and ammunition bunkers were on the north crest. At this point, I should note that I had only four Howitzers rather than the normal six. Each of the Division Artillery Direct Support Battalions provided three Howitzers for a provisional artillery unit providing fire support for the division base camp at Pleiku. I also had one Howitzer in Ordnance as its tube life (the number of allowable rounds fired) had been reached while supporting the initial attack on Hill 1338.

Just before dark on our day of occupation, about half a dozen mortar rounds impacted in the jungle at the base of the north side of the hill. The following evening incoming rounds began to fall on our position. Immediately we commenced our final protective fires. My Cannoneers were superb, remaining at their posts, continuing to fire in spite of the intense shelling—an estimated 75 82mm Mortar rounds fell on or near our position. Reinforcing Fires from an 8-inch, Towed, Battery to our north came to our aide. Fortunately, no ground attack ensued.

By the grace of God, the narrowness of Hill 530, and probability error, our casualties were two KIA and fourteen wounded (none with life or limb threatening injuries). The results could certainly have been much worse. You could hear rounds whizzing overhead and impacting on the south slope – it is the ones you do not hear that get you!

Daylight revealed the physical damage. One Howitzer had both trails bent outward from the concussion of mortar rounds impacting about a foot away from each trail. All four Howitzers had flat tires. This was no small problem. Artillerymen know that the tires (wheels) of the M101 Howitzer are an essential part of the firing platform. We had no capability to repair the flat tires so it was necessary to have complete tires and wheels airlifted to us.

We would remain on Hill 530 without further incident for several days before being airlifted to another inhospitable hilltop further west. That is another story altogether.

Dak To was a significant battle of the Vietnam War. Besides the 4th ID, the 173rd Airborne Brigade and two 1st Cavalry Division units participated. There were nearly 300 killed and 1,200 wounded over the course of a month.

I dedicate this story to the memory of Artillerymen SSG Robert Hackett, SSG Edward Harper, and SGT Donald Mikulecky who gave their all; to SP4 Timothy Nafe and SP4 Robert Butler who survived this battle but would give their lives in the coming months; and to the survivors of Bravo Battery who fought valiantly.

During our B/HSB Reunion in Nashville, October 2016, PFC Louie "Pee Wee" Campbell related some of his personal experiences when he served in B Battery during the Battle of Dak To. He has since followed up with a letter dated 3 Mar 17 which expanded on these events that he was willing to share.

Some of them are humorous, such as this one. "We moved to Dak To sometime around November 1, 1967. We went past Dak To about five miles to an open field next to the highway we set up our guns and started digging. Well, I needed a break and our latrine was not finished so I told the gun Sergeant about it, and he said to take my rifle and toilet paper to the woods and let the people on guard duty know I was out there. I left the safety of the fire base and headed out about 20 yards or so to do my thing. Here I was sitting out there and all of a sudden I heard noises behind me, and the first thing I thought of was, here I am going to get shot by Charlie with my pants down! I put my hand on my M- 16 and kicked the safety off as I turned my head to see a big banana tree with something wrapped around the inside. It had two big eyes and a very big head! All I said was "Oh Shit, a snake, a very BIG snake, looking at me!" I thought about shooting it, but then the guys in the fire base would most likely shoot ME! Well, you can guess what I did---I just stood up and got the hell out of there before I got eaten. Would you believe nobody believed me about the snake and they all laughed about it. Now I can laugh too! A few days later we moved again back to Dak To air strip where we set up for the Battle of Dak To, Hill 1338. The 173rd Airborne Brigade attacked Hill 1338 and we started a mission to support them that lasted all day and night. We ran out of ammo sometime during the night. More ammo was brought into Dak To from Pleiku by truck. I'm glad I did not drive that road after night. When morning broke choppers were bringing wounded down, and KIA in sling loads under the helicopters. That is something I will never forget.

After 1338 was taken, guess who got their gun battery assigned up there—B Battery! We had a Chaplain come to the battery, and to me it was like our Last Rites. What were we getting into?

"On 25 Nov 67 at the FSB on Hill 530 I was a cannoneer on gun #3, led by Chief of Section Staff Sergeant Hackett. It was sometime in the afternoon and I was sitting up against FDC (CONEX) with two other guys in the section. We were supposed to be moving to a higher hill. All of a sudden a mortar round landed about 10 feet from us, and you could hear the shrapnel bouncing off FDC. For some reason not one of us was hit. All three of us headed quickly to the gun positions hollering "Incoming"! It seemed like those mortar rounds were following me---good thing I was young and fast! I had to run the whole length of the battery area to get to my gun. As the mortar rounds landed SSG Hackett was KIA. The story I was told he was looking for me and was in another bunker next to where he was hit. As the night went on every gun was hit, almost direct, and all had flat tires so we could not shift the guns. I don't know how many mortar rounds landed inside the parapet of my gun, but it lasted all night. I was told at least 100 rounds landed on our gun sections. We had wounded to get out somehow. We knew "Charlie" was going to overrun us, or try, and we were ready as we could be, M-16's loaded. But when the sun came up, "Charlie" was gone. We got our ass kicked that night but were determined to come back and do our job! Whoever that little bastard was calling in those mortars knew what he was doing. We lost two KIA and at least 15 WIA that night".

The NVA operational strategy at Dak To was to capture the Special Forces Camp at Ben Het and the airfield at Dak To. This would give them access to Highway 512 running from the Laos and Cambodia border near the Ho Chi Minh Trail, to the Dak To airfield, and beyond to the village of Than Cahn. Than Cahn was next to the major north-south Highway 14 which led south through Kontum and then to Pleiku. A major defeat of US and ARVN Forces at Dak To and the surrounding hills would give the NVA a decided advantage in the II Corps area for the planned Tet Offensive end of January 1968.

Smoke from burning and exploding Dak To Ammunition Dump, SGT Charles Snyder photo.

Engineer Dump Truck severely damaged by huge explosion at Dak To Ammunition Dump, SGT Charles Snyder photo.

The NVA were masters at coordinated attacks. While Mike George and his brave Soldiers were successfully supporting the 3/12th Infantry taking Hill 1338 the NVA attacked the air field at Dak To. Michael Eggleston writes in his book "On the night of 12 November, the NVA launched the first of what would become a series of rocket attacks on Dak To airfield. On 15 November three C-130 transport aircraft were in the turn-around area of the runway when three rounds landed on them, destroying two. The third C-130 was slightly damaged but was able to take off from the Dak To airstrip. The aircraft, with almost full fuel tanks were burning furiously when additional rounds set ammunition and fuel tanks afire. SGM Charles A. Taylor and "our own" Sidney – "Skid-row" Hegener (Recon & Survey Section) both rushed to the ammo dump in a water truck to attempt to extinguish the fire but were driven back by fire and smaller ordnance cooking off before the "big bang" that evening. SGM Taylor earned a Bronze Star w/ V. PFC Hegener earned an ARCOM w/V. Shelling continued into the night, when a well-placed enemy round hit a CONNEX full of C-4 (plastic explosive) container, detonating another nearby. This sent a fireball into the sky that could be seen for miles. Steve Stark and other members of the Recon & Survey Section on the perimeter at Dak To said it was a recoilless rifle fire aimed at the ammo dump and not a lucky shot from a mortar or rocket. They said that the NVA had missed with an earlier shot. The "Ricky Rifle" as we called them were pretty precise in the hands of NVA gunners as proven later when CPT Sammons was at Ben Het when he earned his Silver Star and they almost blew him away several times with well-placed shots. One account said that people a mile away were knocked down by the blast. One soldier near the detonation said that the blast lifted him off his feet and tossed him into a bunker. He was not injured. Steve Stark says "I was blown over the cot in our tent as we had not yet gone underground but was not injured. We were about a half-mile away – perhaps 800 meters. Shrapnel rained down all night and into the next morning. We then manned the perimeter along the river." Two craters over forty feet deep were found in the clean-up after the fire. It is said to be the biggest explosion of the Vietnam War and destroyed over 1,100 tons of munitions." LTC Kenney said in his Operational Quarterly Report "In the morning of 15 Nov 67, the Dak To Base Camp received mortar and rocket fire on the east end of the air strip. The attack lasted approximately 15 minutes. It destroyed two C-130 aircraft. The 6th/29th Artillery Radar located the incoming rounds to be coming from the vicinity of (grid) YB 990 176. B Battery,

6th/29th Artillery and General Support Artillery fired the counter mortar program. About 1600 the weapons attack resumed. The ammo dump was hit and rounds from the ammo dump blew for about three hours. Mortars were observed firing from vicinity grid YB 990 176. B Battery, 6th/29th Artillery fired into this area and the mortars were silenced."

The NVA and VC forces had enough by the end of Nov 67. Willbanks writes in his book "During the bitter fighting three PAVN (NVA) regiments were severely crippled, suffering an estimated 1,400 soldiers killed, but allied losses were heavy, too. The ARVN lost 73 men, and 289 Americans were killed during the battles around Dak To; More than 900 American soldiers were wounded." The "extraordinary heroism" of the 6th/29th Artillery and the other US units who fought in the Battle of Dak To was recognized by President Richard Nixon General Order 38, The Presidential Unit Citation (Army) for Extraordinary Heroism, shown here:

General Orders
No. 38

HEADQUARTERS
DEPARTMENT OF THE ARMY

Washington, DC, 20 July 1971

THE PRESIDENTIAL UNIT CITATION (ARMY). Award of the Presidential Unit Citation (Army) by the President of the United States of America to the following units of the Armed Forces of the United States is confirmed in accordance with paragraph 194, AR 672-5-1. The text of the citation, signed by President Richard Nixon on 5 May 1971, reads as follows:

By virtue of the authority vested in me as President of the United States and as Commander-in-Chief of the Armed Forces of the United States, I have today awarded

THE PRESIDENTIAL UNIT CITATION (ARMY)
FOR EXTRAORDINARY HEROISM
TO THE
HEADQUARTERS AND HEADQUARTERS COMPANY
1ST BRIGADE, 4TH INFANTRY DIVISION (FIRST
OAK LEAF CLUSTER)
1ST BATTALION, 8TH INFANTRY
3RD BATTALION, 8TH INFANTRY
3RD BATTALION, 12TH INFANTRY (1ST OAK LEAF CLUSTER)
TROOP C, 2ND SQUADRON, 1ST ARMORED CAVALRY
COMPANY B, 1ST BATTALION, 69TH ARMOR (THIRD OAK
LEAF CLUSTER)
BATTERY C, 6TH BATTALION, 14TH ARTILLERY
BATTERY A, 5TH BATTALION, 16TH ARTILLERY
BATTERY B, 5TH BATTALION, 16TH ARTILLERY
BATTERY D, 5TH BATTALION, 16TH ARTILLERY
6TH BATTALION, 29TH ARTILLERY
BATTERY A, 4TH BATTALION, 42ND ARTILLERY
2ND PLATOON, BATTERY B, 4TH BATTALION, 60TH ARTILLERY
BATTERY B, 1ST BATTALION, 92ND ARTILLERY
52ND AVIATION BATTALION
COMPANY A, 4TH ENGINEER BATTALION
COMPANY D, 704TH MAINTENANCE BATTALION

1ST PLATOON, 4TH MILITARY POLICE COMPANY
COMPANY B, 4TH MEDICAL BATTALION
1ST PLATOON, COMPANY B, 124TH SIGNAL BATTALION
DETACHMENT 1, 374TH RADIO RESEARCH COMPANY
AND
TEAM 1, 4TH MILITARY INTELLIGENCE DETACHMENT
UNITED STATES ARMY

The 1st Brigade, 4th Infantry Division, together with attached and supporting units, distinguished itself by extraordinary heroism in action against a hostile force in Dak To District, Kontum Province, Republic of Vietnam, during the period 29 October 1967 to 30 November 1967. Reacting promptly to intelligence indicating a major enemy offensive attack against Dak To Base and the Central Highlands, the 1st Brigade deployed its forces to block the impending assault. Seizing the initiative by striking the enemy before he could attack, the Brigade with unrelenting courage and dedication fought and defeated a numerically superior enemy during thirty-three days of violent, sustained combat. Vigorously executing a sound tactical plan with a spirit of total dedication to mission accomplishment, the Brigade severely crippled three enemy regiments and drove the entire force from the Central Highlands, inflicting on them devastating losses in personnel, supplies, and equipment. The relentless determination against overwhelming odds, the sustained display of self-sacrifice, the untiring devotion to duty, and the extraordinary heroism of the officers and men of the assigned, attached, and supporting units of the 1st Brigade, 4th Infantry Division, are in keeping with the highest traditions of military service and reflect great upon them and the Armed Forces of the United States.

Any 6th/29th Artillery Soldier assigned to the battalion from 29 October 1967 to 30 November 1967 earned this Presidential Unit Citation as a permanent award and it should be in his permanent military record.

Though the Battle of Dak To had been won by the US and ARVN allies, the NVA did not fade away completely. They were taking a “breather”. The NVA leadership had not abandoned their “Grand Plan” for the Tet Offensive.

Dak To epitomized the Central Highlands of Vietnam. A study of the map shows that Dak To is about 90 kilometers (about 55 miles) north of Pleiku in very mountainous terrain. Significantly, the borders of Vietnam, Cambodia, and Laos all come together in what was known as the “Tri-Border Area”, just 25 kilometers to the west. The Ho Chi Minh Trail, infamous supply line from North Vietnam cut through just to the west of the Tri-Border Area. David Maraniss describes the Ho Chi Minh Trail in his book “They Marched Into Sunlight”. “The Trail, which started as a scraggy walking path wending south along the spine of the Truong Son Mountain Range, had evolved over years into a veritable interstate system of jungle shrouded trails and roads (some, by the fall of 1967, when we were there, allowing motorized transport) that snaked down through North Vietnam, Laos, and Cambodia, with several arteries slicing west to east to supply the VC and NVA units in South Vietnam.” The 105 mm howitzer batteries of the 6/29th Artillery fought the Battle of Dak To from fire support bases, most located on mountain tops. Fire Bases were numbered for identification, and some even took on nicknames such as “The Peanut, Dog Bone, Ski Lodge, or Eagle’s Nest”. Typically, a Highlands firebase was on a mountaintop, occupied by a Direct Support (DS) 105 battery co-located with the supported Infantry Battalion Headquarters

Tactical Operations Center (TOC). The DS battery was organized with five M101 105 mm howitzers. These were WWII and Korean War vintage, and much of the ammunition was dated from the same era. The Army standard organization for a 105-mm battery was six howitzers, but in the 6/29th Artillery each battery (A, B, and C) provided one howitzer to be left back at Camp Enari to provide fire support for Base Camp Defense. "B" Battery, 6/29th Artillery was assigned the mission of providing Direct Support to the 3rd Battalion, 12th Infantry. On the firebase the 3/12th Infantry was organized with four maneuver companies, A, B, C, and D company. One company, on a rotational basis, provided perimeter security for the fire base, freeing the remaining three companies for operational missions. The firing batteries had their own TOC, known as the Fire Direction Center, or FDC. The FDC was housed in a modified CONEX container, normally not far from the supported Battalion TOC. The CONEX housed two key pieces of equipment for the FDC: The Field Artillery Digital Analog Computer (FADAC), and radios.

Well developed B Battery howitzer parapet on Hill 1338 overlooking Dak To. Sandbagged protected personnel and ammunition bunkers in place, SGT Nick Gillen photo.

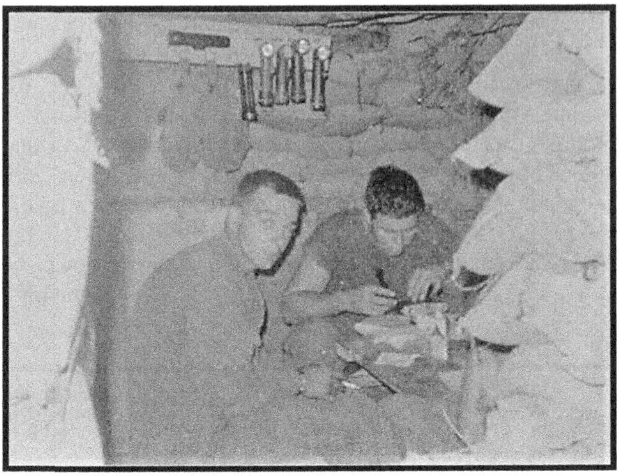

SGT Nick Gillen (L) and CPL Donald Fink at home in their protected personnel bunker within easy access to their howitzer, Hill 1338. SGT Nick Gillen photo.

The howitzers were usually positioned in a circle about 25 meters apart surrounded by a protective parapet made with sandbags filled with dirt, a complete circle enclosing the howitzer. The requirement was to be able to fire in any direction quickly. Each howitzer had a section (crew) of six to seven Soldiers, and their living quarters was a sandbag bunker built tangent to the parapet circle, with a blast protected entrance from the bunker to the howitzer. Another bunker housed the ammunition, offset from the crew bunker to allow maximum protection for the cannoneers in the event the ammunition bunker was hit by incoming rounds and exploded. A key building material for the bunkers was Perforated Steel Planking, known as PSP. PSP planks were about 2 feet wide, and 10 feet long. They were designed to be used to quickly construct temporary landing strips and pads for aircraft. PSP proved indispensable for building bunkers of every description, and they were perfect for building parapets for the howitzers. They became the ceiling of the bunkers upon which three layers of sandbags were stacked. The Division had a Standing Operating Procedure (SOP) that Soldiers and ammunition must be protected by a minimum of three layers of sandbags when moving into a new position before any stand down was permitted. The howitzer, protected by a parapet and two bunkers, was the "home" of the cannoneers on a "Highlands" fire base.

This "home" for the cannoneers was not without unique amenities to make life on a combat fire base a little more pleasant, healthy, and safe. Every Artilleryman in the Battalion was issued a mosquito net to hang over his bunk/air mattress whenever he bedded down. Only our Redlegs in the FO parties were without them when they were out in the "boonies" with the Infantry. The nets were to protect against bites from malaria infected mosquitos. Further protection against malaria was the malaria tablets, affectionately known as "horse pills" because they were so large, were to be taken every day before evening chow. At the same time, sleeves on the jungle fatigues were to be rolled down for the night, the rationale being that mosquitos were more active at night. I think many of us suffered a good case of the "runs" for a week or so after we started taking the malaria pills.

Australian Showerhead supported by A-Frame (center in photo) on Hill 1338 overlooking Dak To. SGT Nick Gillen photo.

On the fire bases there was always a water trailer (known as a "water buffalo") which provided water to the Soldiers for filling their canteens, shaving, and to the mess section for cooking. A really neat invention called an Australian Showerhead provided a way for our Soldiers to take a shower on the fire base. The Australian Showerhead was a canvas bucket with a handle. On the bottom was a showerhead which screwed in and out. When full of water and the showerhead is screwed in, stopping the water flow, the bucket can be hung up, and then the showerhead can be screwed out to let the water sprinkle out as in a shower. For a real treat, the water can be heated with the immersion heater in the mess section!

We had a great mess section on B Battery's fire bases. Our Mess "Sergeant" was PFC Chris Wright and we called him "Pineapple" because he was from Hawaii. Everybody loved Pineapple because he was always smiling and upbeat, with an encouraging word for all who came through his mess tent every day. He and his assistant, PFC Knight, would have fresh coffee, scrambled eggs, and bacon every morning. The eggs were powdered, and the bacon canned and dehydrated. Somehow Pineapple made it taste good. The evening meal was also reconstituted, usually mashed or scalloped potatoes, meat, and a vegetable. Pineapple and his crew worked miracles with this stuff! The noon meal was easy for the mess section—C rations! Usually Pineapple would heat the entre' of the C rations in hot water with the immersion heater. Everybody had their favorite C ration meal. Mine was beans and franks, with peanut butter and crackers. I believe most would agree that ham and limas were the least favorite among our soldiers (Notable exception—1LT Pat McNulty explains in his FO description). The little sundry pack in the C Ration box contained a very important item—a small packet of toilet paper which came in handy many times! C rats lent themselves to a lot of trades. I did not smoke, so I always traded my three-cigarette pack for—peanut butter and crackers! One of the most important tools in Vietnam was the P-38 can opener required to open the C ration cans. Most Soldiers attached them to their dog tag chain. I still have mine!

Sundry Packs came in large sizes too. These were welcome "goody" boxes that provided a wide variety of treats that made life a bit more pleasant for Soldiers. Among the more useful were shaving items such as shaving cream and blades. I believe there was soap in there too. There was candy too, and my favorite was the Charms jelly candy which came in a variety pack of flavors. My memory is fuzzy after all the years, but I think playing cards (Bicycle), and stationery for letter writing and ball point pens were included too. None of these necessities would have been available to our Soldiers had it not been for the Sundry Packs.

Top and bottom photos: Red Cross Girl (Donut Dolly), name may be Mavis, visiting Fire Base 25 near Dak To. Photo source unknown.

Nothing makes a Soldier on a remote fire base feel more at home than a cold beer and a visit by pretty American Girls. The pretty girls were the Red Cross Girls, affectionately known as "Donut

Dollies". These young ladies were about the same age as our young Soldiers, visiting all units in the 4th ID in the fire bases and at Camp Enari. They were like the "round eye" girls 'next door' that the Soldiers remembered from home. They would sometimes play "silly" games with the GI's to get them involved, but the most value of their visits was reminding the Soldiers of what they were fighting for back home. Not to be overlooked is that they were also risking their lives in a dangerous combat zone. I remember a Red Cross Girl visiting us on FSB 25 which was being attacked virtually daily in the Jan-Feb '68 timeframe. No mortar rounds landed during her visit. I have great respect for what the Donut Dollies did for the morale of all of us. As for the beer, I believe the daily limit was two beers, 3.2 alcohol. I recall Ballentine was a pretty common brand, in rusty cans. But who cared—it was cold and wet!

B BATTERY AT DAK TO (AFTER THE BATTLE)

CH-54 Flying Crane bringing in heavy load to Fire Base known as "Christmas Hill" west of Dak To near Cambodian Border. Note purple smoke marking LZ. SP4 Sam Farris photo.

Church Service being conducted on Christmas Hill. 1LT Bill Davis photo

Blindfolded NVA soldier being questioned by Vietnamese interpreter on Christmas Hill, SSG Donald "Buck" Buchanan, Howitzer Section Chief, looks on. SP4 Sam Farris photo

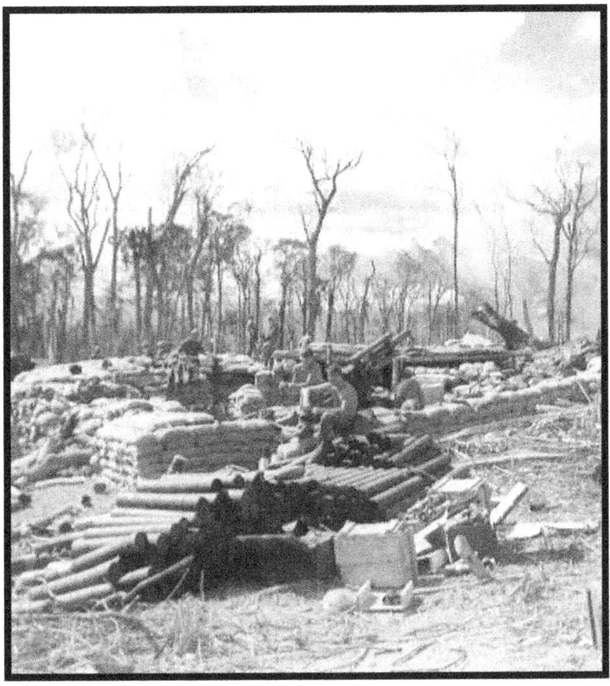

Three photos above, all appear to be taken on Christmas Hill west of Dak To, near Cambodian Border. Top photo: L-R SSG Larry Kidwell, Chief of Section, SP4 Larry Manny, loading round in preparation to fire, Unknown Soldier performing Gunner duties on panoramic telescope, Possibly SP4 Robert Butler on R performing Assistant Gunner duties. Middle photo: Well developed howitzer parapet, Gun Section name unknown. Bottom photo: L-R SP4 Manny toting M-60 Machine Gun, other Soldiers unknown. SP4 Larry Manny photos.

On 29 Nov 67 B Battery moved again, further to the west, closer to the Laos/Cambodia Borders. It was Hill 1030, but the Soldiers dubbed it "Christmas Hill" because the battery was there during Christmas, 1967. It was here that I assumed command of B Battery from Captain Mike George on 7 Dec 67.

When I assumed command of B Battery I inherited a battery of battle hardened and skilled veterans who had just acquitted themselves with honor in one of the key battles of Vietnam.

I quickly learned to respect and value the Officers and NCOs in the battery. The battery was then positioned on Hill 1030 west of Dak To, within sight of Hill 875, signature site of the Battle of Dak To less than a couple of weeks before. The Battle of Hill 875 is popularly known as the 173rd Airborne Brigade Battle, but the "rest of the story" is that the 4th Infantry Division was a major player in the ultimate victory in that key battle in Dak To. Hill 875 was just north of Hill 1030, or "Christmas Hill". Hill 875 showed the effects of the fierce battle. The mountain top was completely denuded of any vegetation by the intense bombing and artillery fire employed during the battle.

The "Christmas Hill" experience was very instructional for me. B Battery shared the fire base with our supported Infantry Battalion, the 3rd Battalion, 12th Infantry. Every evening the battalion held a meeting at their Tactical Operations Center (TOC) to brief on events of the day, and to discuss details of the operational plans for the next several days. The meeting was called by the Battalion Commander, Lieutenant Colonel (LTC) Jamie Hendrix, and run by the Battalion Operations Officer (S-3), Major Carl Stiner. Other staff officers were present, including Captain Hill, the Battalion Artillery Liaison Officer from the 6/29th Artillery. "Grizzly", the radio nick name of LTC Hendrix, always wanted his Direct Support Battery Commander present at the evening meetings to insure we were in sync with his battle plan. This battle plan was always spelled out by Major Stiner and included the plan for fire support provided by B Battery, and other reinforcing artillery and close air support coordinated by CPT Hill, the Artillery Liaison Officer. These meetings provided a critical coordination between the fire support means and the Infantry companies with their Artillery Forward Observers maneuvering through the jungle seeking to make contact with the enemy.

The time on Christmas Hill was busy, though nowhere near the level B Battery saw during the Battle of Dak To. Fire missions were requested by our Forward Observer teams to support the Infantry during their "fire fights". These were on 1, 3, 5, 11, 12, 13, 14, 15, 17, 24, 25, and 26 Dec 67. Most missions were in support of 3rd/12th Infantry; some for the 3rd/8th Infantry. There was supposed to be a Christmas Truce—the NVA broke it, and B Battery fired on Christmas Day. I believe the 3rd/12th Infantry had a Lieutenant KIA on Christmas Day. That got to me! One of the more unusual missions was directed by the 6/29th Battalion TOC at Dak To. We were provided an overlay with several targets plotted on it. The targets were located in Cambodia and Laos, across the Vietnam international border, the Tri-Border area. Presumably these targets were intended to interdict the Ho Chi Minh Trail which passed through that area. There was something exciting about firing at targets "across the border"!

Neither the Military Leadership in Vietnam, nor our families and supporters back home forgot about the importance of celebrating the Christmas Season while the fighting continued. Bob Hope brought his Christmas Show to Vietnam putting on a performance for the 4th ID Soldiers at Camp Enari. One of our B Battery Soldiers, PFC Louie "Pee Wee" Campbell was selected to get a trip to base camp to see the show. Louie was a cannoneer on one of our howitzers (The Brute). When he left home for Vietnam, Louie purposely did not tell his grandmother that he was going to Vietnam because he did not want her to worry. The Bob Hope Christmas Shows were taped and shown back home in the USA during the Christmas season. They were very popular prime time shows. Louie's grandmother was watching the show which included clips from Camp Enari. She saw a familiar face in the crowd of Soldiers and was reputed to exclaim "That's my Louie"! Just goes to show you can never fool Grandma!

Prayer for Christmas

Our Father in Heaven, we give thee thanks for the gift of thy Son Jesus Christ, our Savior, the Prince of Peace and Lord of life. May thy gift at Bethlehem [illegible] of thy message: "Peace on Earth, Good Will Toward Men." [illegible] and help us to [illegible] thy peace [illegible] understanding. We pray that the spirit of Christmas will be shared by [illegible] memory of Christmas. May the [illegible] AMEN.

Commander's Message

Throughout the Christian world, the Christmas season is a time of joy and spiritual inspiration. Despite separation from our families and the hardships imposed by war, those of us in Vietnam will still share the traditional Christmas spirit this year. We can enjoy the spiritual satisfaction that comes from giving. As fighting representatives of the Free World, our gift is the help we give the Vietnamese people to secure their independence, their individual safety, and their future freedom. Each of you gives a part of this gift and deserves the satisfaction of having increased the [illegible] of [illegible] the true Christmas spirit.

My best wishes to each of you and your families for this Christmas season. May you enjoy good fortune during the coming year.

W. C. WESTMORELAND
General, U.S. Army
Commanding

Christmas message to the troops from General Westmorland, Commanding General US Forces, Vietnam. Also included inside the brochure was the menu for Christmas Day Dinner. Provided by SP4 Larry Manny

There were other fire missions directed by the 6/29th Battalion TOC. These missions were called "Harassing and Interdiction" (H+I) fires. These missions were almost always fired at night, and their purpose was to keep the enemy off guard and uneasy while moving around at night. The goal was to drop an unexpected volley on the enemy, demoralizing them. Target grid locations were provided by the 6/29th TOC. Co-author Steve Stark was part of the team at 1st Bde HQ who developed selected targets for 6/29th and many other Artillery units. There was a time sequence given to the targets, attempting to match the estimated time window when the enemy would be there. On the fire base, there was something reassuring about hearing the H+I missions firing throughout the night. It was a sound you became accustomed to hearing, knowing the rounds were going out, not coming in! I found myself saying "All is Well". As a new Battery Commander, I quickly learned that life on the fire bases was simple, basic, and "close to the earth". I described earlier the living bunkers for the gun sections, part of the howitzer parapet. All of our Soldiers had protected bunkers as their living quarters. Even in the austere field living conditions, our resourceful Soldiers found ways to make life on a firebase acceptable. On 29 Dec 67 B Battery was airlifted to Dak To Base Camp. It was my first move.

GRIZZLY'S SKI LODGE

All photos Grizzly's Ski Lodge, one mile high and 25 kilometers northeast of Dak To. SP4 John Nicolson photos
Top: CH-54 Flying Crane lifting in 155 Howitzers from 1/92 Artillery Middle: 3rd 12th Infantry TOC
Bottom: Crashed CH-47 Chinook awaiting extraction by CH-54 Flying Crane.

After three days of rest and refit at Dak To, B Battery was airlifted to a fire base 25 kilometers to the northeast of Dak To on 2 Jan 68. I don't believe it ever had a number. Nor do I remember what the enemy intelligence reports were that drew us to that location. One thing I do remember is, the mountain top we were on was one mile high! It was literally above the clouds. Another thing I remember for sure---it was cold as hell up there! And, this was the first and only FSB where the perimeter security was provided by ARVN soldiers. I would have been nervous about that, except that we were so high up, and the hill was so steep, I wasn't too worried about it. Before I go forward any more, I need to explain the name "Grizzly's Ski Lodge". Pretty easy. We all (6th/29th Arty and 3rd/12th Inf) were freezing our asses off on that mile-high mountain top. From that came "Grizzly's Ski Lodge"! Occupying this fire base was one of the more bizarre experiences B Battery experienced in Vietnam. It was very cold on this mountain top, especially at night. This is my guess, but I suspect the temperature hovered around 40 F degrees at night. We were not acclimated to these temperatures in Vietnam, so we requested an emergency supply of field jackets and gloves! I was very favorably impressed that our Battalion S-4, Captain Richard Wright, was able to get the jackets and gloves for everybody in the battery out to the fire base in a day or so. Grizzly's Ski Lodge was one of the larger fire bases B Battery occupied. We made room for a firing battery from the 1st/92nd Arty, a 155 towed (M-114) howitzer battery. They were airlifted in by CH-54 Flying Crane Helicopters. We didn't do a whole lot if firing while we were at the "Lodge". The 3rd/12th Inf had a few contacts, all with the VC. We did have some excitement, though. A Chinook helicopter with a sling load—I believe it was a water "buffalo", was coming in to the fire base to drop the load, but it was coming in too low. The water buffalo dragged on the ground, pulling the Chinook over on its side, and on to the ground. Miraculously no one was hurt by the flailing double chopper blades. The damaged Chinook required another visit by a Flying Crane to lift it out and take it back for repair. On 14 Jan 68, after almost two weeks of relative inactivity, B Battery was airlifted to FSB 25. The inactivity was about to end.

FIRE SUPPORT BASE 25--PRELUDE TO THE TET OFFENSIVE

For virtually the next two months, B Battery would affirm the "Baptism of Fire" it earned for "Extraordinary Heroism" during the Battle of Dak To. For my description of B Battery's combat experience at FSB 25 I will rely on my personal recollections, daily log reports from 6th/29th Artillery, and Quarterly Operational Reports from two Battalion Commanders, LTC's Kenney and Akers, cited earlier in this history. FSB 25, northwest of the Dak To base and airfield, and northeast of the Special Forces Camp at Ben Het, also designated as FSB 12. From FSB 25 B Battery could easily range and provide fire support to both locations and had good observation and coverage of Highway 512 which connected with the Laotian and Cambodian border, snaked east past Ben Het, Dak To air airfield, and then to Tan Canh, where it connected with the major north/south Highway 14. Thus, FSB 25 was critical military terrain for the 4th ID in Dak To. As we would see in the next two months, the NVA considered it critical military terrain as well!

Fire Support Base 25 approached by helicopter from Dak To to the southeast. Note infantry fighting position bunkers down from the military crest. Source of photo unknown.

View looking west from FSB 25 towards Ben Het and Cambodia-Laos border. SP4 Sam Farris photo.

3rd Bn 12th Infantry TOC on FSB 25. B Battery FDC TOC was right next door. SP4 John Nicolson performed Liaison duties here. SP4 John Nicolson photo

On 14 Jan 68 B Battery was airlifted from Grizzly's Ski Lodge to FSB 25. While the battery was being deposited on the fire base, the enemy greeted us with mortar fire. I was there with the Advance Party, and it was my first experience with a mortar attack. Since we had just arrived, there were no protective bunkers prepared. I learned quickly what it meant to "get as close to the ground as you could"! About the same time we were receiving "incoming", the Special Forces Camp at Ben Het received about 40 rounds of 82mm mortars. Looking back, I realize the enemy was firing at FSB 25 to suppress our ability to provide fire support to Ben Het. It didn't work. B Battery fired a counter mortar program on grids provided by a Q-4 radar located at Ben Het. The An/MPQ-4 Radar belonging to the Battalion moved to Dak To on 31 October 1967. The set was initially surveyed in by the Survey Section of the 6th Bn, 29th Artillery at coordinates YB 00432 21712. The mission of the Radar was twofold: provide counter-battery and counter-mortar capability by determining grid locations of enemy artillery, rocket, and mortar locations, and high-burst registration capability for our firing batteries. The Radar Section was led by Warrant Officer (WO2) Ashby Owens and his small crew of radar specialists. Battalion Commander LTC Donald Kenney reports in his Operational Quarterly Report ending 31 January 1968 the 6th/29th Artillery Q-4 Radar section conducted 25 high burst registrations, located 21 possible enemy mortar locations, and 1 rocket location. The Q-4 Radar was an essential tool for the Battalion at Dak To. In addition to its counter battery/mortar/rocket capability, the Radar was a key player in enhancing the accuracy and safety of the fires of the Battalion's three firing batteries. The dense triple canopy jungle prevented the firing batteries from conducting "impact" (on the ground) registrations—the rounds simply could not penetrate the jungle. The "high burst" registration, made possible by lifting the procedure above the jungle using time fuses bursting in the air, could only be accomplished using the Q-4 Radar measuring the location of those air bursts. Registering the firing batteries on a regular basis, preferably every other day, and most importantly, after a move, was a standard artillery gunnery procedure to determine corrections to firing data to increase accuracy and safety of Artillery fire being provided in support of the Infantry. The Q-4 Radar was very versatile in the skilled hands of operators in the Battalion Radar Section. It could also be used to adjust "friendly" Artillery fire on the target. The Radar Section was equally adept at determining the impact location of "outgoing" rounds as well as the more usual "incoming" rounds. While on FSB 25, I recall using the Q-4 Radar to conduct "high burst" registrations, and also to adjust fire at targets near the Special Forces Camp at Ben Het.

As was standard procedure, we shared FSB 25 with our supported battalion, the $3^{rd}/12^{th}$ Inf. They were organized with four rifle companies, one of which provided perimeter security around the fire base on a rotating basis. The fire base was tight, but not cramped. The Infantry Battalion TOC was not more than ten feet from our FDC CONEX. As per Standard Operating Procedure (SOP) we prepared our positions for our five howitzers and the FDC using PSP and sandbags to provide a minimum of three sandbag layers of overhead cover for every Soldier before going to sleep. It was late into the night before we could think about sleep, but we were in position and ready for the long haul.

On 15 and 16 Jan 68, FSB 25 received incoming rounds, while Ben Het received incoming on 16 and 17 Jan 68. We were now seeing an NVA pattern of operational tactics—capture Ben Het with coordinated attacks on the Special Forces Camp and FSB 25 while pinning down the Infantry and supporting Artillery there. The $3^{rd}/12^{th}$ Inf and B $6^{th}/29^{th}$ Arty were not about to let this happen.

"On 19 Jan 68, 0845, B Company $3^{rd}/12^{th}$ Inf was ordered to search and clear the ridgeline northwest of FSB 25 with primary objective being Hill 800, known as the "Peanut" (near FSB 25), nick-named so because it resembled the shape of a peanut when viewed on a topographic map. B Company was attacked from the objective area by a 60-man NVA force from three directions. B Company formed a perimeter and attempted to recover the dead and wounded. The company was ordered to withdraw 200 meters to allow more effective artillery and mortar fire to be brought to bear upon the enemy. The Forward Observer Team making this happen was First Lieutenant Patrick McNulty and his Radio Telephone Operator, Specialist 4th Class Timothy M. Nafe. During this action, B Company sustained 35 casualties including the Company Commander. While calling for fire support, Specialist Nafe was Killed in Action. Early in the morning of 20 Jan 68 the enemy again used coordinated attacks on FSB 25 and Ben Het. Mortar rounds impacted near Specialist 4th Class Robert E. Butler who was killed in action, and wounded Specialist 4th Class Larry G. Manny, Specialist 4th Class Lester W. Loo, Sergeant (E-5) James E. Hook, and Private First-Class David Henderson, all cannoneers on the howitzers. Specialist Manny was severely wounded, requiring "Dust Off". Larry required a full year of recovery, finally discharged from Valley Forge, Pennsylvania Medical Center with a Medical Retirement from the US Army. Specialist 4th Class James A. (Jimmy) Williams related to me in a conversation with him on 19 Feb 2017 his personal experience with the mortar attack. He was a fellow cannoneer with Larry Manny on Chief of Section Staff Sergeant Charles L. (Larry) Kidwell's howitzer. Two mortar rounds came in. One severely wounded Larry Manny. The second round killed Robert Butler. The blast of these rounds was so loud it severely damaged Jimmy Williams' hearing. A few days later Jimmy sustained further ear damage when SSG Graves, a neighboring Section Chief fired his howitzer without warning Jimmy. On medical orders he was taken off the howitzer and sent to Dak To. There he was given the duty to prepare and hook up ammunition sling loads to Chinooks headed to the various fire bases resupplying the firing batteries surrounding Dak To.

According to the 6th/29th Arty Daily Log (DA Form 1594) for 20 Jan 68, at 3:00 AM in the morning Captain William A. Sammons, 6th/29th Battalion LNO to 1st Bde HQ at Dak To, and detailed to Ben Het Special Forces Camp to coordinate fire support and security, reported NVA close to the wire on the perimeter. "Spooky" was called to provide illumination. Air strikes using napalm followed. NVA attack repulsed.

An After Action Review (AAR) from 3/12 Infantry dated 20 Jan 68 reported "Seven airstrikes were called in on enemy positions. Throughout these airstrikes, the NVA continued to direct a heavy volume of automatic weapons fire at the Air Force fighters and periodically placed sniper fire into the firebase. The enemy force was determined to be a battalion-sized element of the 66th NVA Regiment, capable of sitting through the airstrikes and artillery fires in relative safety while trail watchers observed FSB 25. The next day, 21 Jan 68, five TAC airstrikes were placed on and around Hill 800, the "Peanut". During each strike, aircraft received heavy fire from at least four machine gun positions along the ridge." I remember watching the F-4 Phantoms swooping in just to the north of the fire base. As they approached the target, I could hear the NVA gunners firing their Anti-Aircraft guns (probably .51 caliber Russian design) at the Phantoms. Then the fighters would drop their bombs, and you could hear the bomb fragments and debris whizzing past overhead. Glad the NVA was the target, not us! Again, Ben Het received a weapons attack. I had an interesting recent conversation with our Battery Executive Officer, 1LT Ravin (Ray) Carpenter, about all of the action centered around FSB 25. He reminded me of two incidents that took place toward the end of January '68 just before the Tet Offensive. I remember both. In one instance, the relative quiet of mid-day on the fire base was shattered by an ear-splitting blast. Everybody on the fire base, Artillery and Infantry alike, were shaken by the huge bang. Our first thought was that it was incoming. We soon realized it had come from the direction of U.S. artillery near Dak To. A further check determined that it was fired by a 155 mm battery. The round hit a large tree stump, and by some miracle, nobody was injured. Ray had just finished rechecking the "lay" of the battery with the aiming circle and had moved back into his protected bunker which served as his XO's post. When the round exploded, it sheared off the aiming circle mounted on the circle's tripod. Ray had been standing next to the aiming circle just moments before. It was not his time to go!

In the other incident, Ray noticed that the 3rd/12th Infantry, with whom we shared FSB 25, had an ammunition bunker which held 81mm mortar rounds, hand grenades, and small arms ammunition near the fire base perimeter. The bunker was not protected by overhead cover—the ammo was in the open. Concerned about the exposed ammo, Ray told the 3rd/12th Battalion Commander, LTC Jamie Hendrix (Grizzly), "Sir, you need to get that ammo bunker under cover so it won't be exposed to enemy fire, like we do here in B Battery". Grizzly agreed, but before his Soldiers reacted to get the bunker covered, the exposed ammo in the bunker was hit by incoming enemy fire and the bunker was destroyed before the Soldiers could put the fire out that was cooking off the rounds in the bunker. 1LT Howard "Tappy" McEwen, our Fire Direction Officer, once shared with me personal notes he included in a scrap book he created to cover his Vietnam experience. He said the rounds were cooking off for two days, and over 4000 rounds were blown up!

During this period of heavy combat action on FSB 25 an interesting twist happened most likely sometime the last two weeks in January 1968. "Hanoi Hannah" came on the radio and warned the Soldiers of the 3rd/12th not to leave FSB 25 for their next "search and destroy" mission. "Hanoi Hannah" was an NVA propaganda lady patterned after the infamous "Tokyo Rose" for Japan in

WWII. I heard "Hanoi Hannah's" broadcast—can't remember if it was over our FM tactical radios, or overriding Armed Forces Vietnam Radio (AFVN), the station run by the US Military for the troops in Vietnam. It was weird to hear her addressing the Soldiers on FSB 25, warning them if they ventured out they would be sent home in a body bag!

The 6th/29th Arty Daily Log for 27 Jan 68 recorded that at about 7:00 AM, both FSB 25 and Ben Het received incoming mortar rounds. Their firing locations were plotted by Q-4 Radar from the 237th Radar Detachment and counter mortar was fired. Three B Battery Soldiers were Wounded in Action (WIA) by the mortar rounds: Sergeant Nicholas J. (Nick) Gillen (Assistant Howitzer Section Chief), Private First-Class Chris D. (Pineapple) Wright, our Mess "Sergeant", and Specialist 4th Class Salvador M. Sanchez (Howitzer cannoneer). One funny story came out of this serious incident. "Pineapple" Wright was from Hawaii, and of Asian descent. Apparently when the three arrived at the Aid Station at Dak To, the Medics asked Pineapple if he was an enemy soldier! Shortly thereafter the 6th ARVN ABN BN, working together with the 3rd/12th Inf to finally wrest the "Peanut" from the NVA, was attacked by a battalion sized NVA unit. B Battery responded with quick fire support. The prelude to the Tet Offensive was about to come to an abrupt end. The shock of the 1968 Tet Offensive was upon us.

FIRE SUPPORT BASE 25—TET OFFENSIVE 1968

The Grand Strategic Plan to achieve victory over the ARVN and US Forces in South Vietnam, conceived by the NVA leadership and planned by General Giap in mid-1967, was about to begin. Willbanks writes "Vietnam's Tet holiday, which marks the start of the lunar new year, is a time not only of revelry celebrated with feasts and fireworks, but also of worship at the family alter for revered family ancestors. For several days the entire countryside is on the move as people visit their ancestral homes, and all businesses, including the business of war, comes to a halt. Before 1968, both sides of the war had observed Tet cease-fires over the holiday. Therefore, The North Vietnamese reasoned that half of the South Vietnamese army and national police would be on leave when Tet 1968 (New Year of the Monkey) began on 31 Jan 68, and Saigon would be unprepared for a countrywide attack."

3/12th Inf AAR, dated 29 Jan 68. "The Tet 68 Truce began at 6:00 PM on 29 Jan 68. At 7:46 PM the NVA violated the short-lived truce by launching a sizable mortar attack against the "Peanut", where 3rd/12th Inf was located, and nearby Hill 900 (1st/8th Inf) was probed throughout the night. Late afternoon on 30 Jan 68, FSB 25 received more than 60 rounds of 82mm and 120mm mortar fire which resulted in eight infantry casualties. The 120mm mortars posed a real threat to us in the battery. Our bunkers were protected with the SOP three layers of sandbags laid on top of single sheets of PSP. They could protect against 82mm mortars, but not the larger 120mm mortars, larger than our 105mm artillery shells. Next day, late afternoon 31 Jan 68, C Company, 3rd/12th Inf, who was providing perimeter security at FSB 25, reported an ignited trip flare approximately 150 meters to the northwest in the direction of the "Peanut". About 6:30 PM, an eight-round volley of 82mm mortar fire struck the fire base. During the next hour and four minutes, more than 150 rounds of 82mm and 120mm mortar fire, in 6-8 round volleys, impacted inside the perimeter, causing 11 infantry casualties." The thing I most remember about this mortar barrage was the intense, prolonged noise of the rounds impacting. We knew something big was about to happen. By this time, it was starting to get dark. "Nine minutes after the mortar barrage ended, enemy soldiers were observed approximately 50 meters from the perimeter bunkers. Grenades, M-79s, and artillery and mortar final defensive fires, including those fired by B Battery, were continuously fired for one hour and 15 minutes. About 9:00 PM the NVA fired 25 RPG-2(B-40) rockets, in volleys, at the perimeter and, in an attempt to breach the bunker line followed this with a ground assault supported by automatic weapons." It was now very clear that the NVA were planning to overrun FSB 25! Company C countered this attack with a heavy volume of automatic weapons fire, while 3rd/12th Inf mortar fire was adjusted to the edge of the perimeter. I distinctly remember the 50-caliber machine gun the Infantry had set up on the northwest corner of the perimeter. It was firing constantly, with that classic loud "rat-tat-tat" sound only a 50-caliber could make. It was reassuring!

By now I was in B Battery's FDC TOC on the landline phone with LTC Hendrix, CO 3rd/12th Inf, in his nearby battalion TOC. He was also in landline communication with C Company Commander, Captain Kresefski, who was responsible for perimeter security and reporting to LTC Hendrix where enemy activity was most threatening. LTC Hendrix wanted to bring the direct firepower of B Battery's 105 mm howitzers to bear on the attacking NVA. He directed me to stay on the line and have the battery prepare to fire "beehive" rounds. 1LT Howard D. "Tappy"

McEwen, B Battery Fire Direction Officer, voluntarily left the safety of the FDC bunker and went out among the howitzers, under heavy fire, to assist in controlling the direct fire. A pattern of Infantry-Artillery teamwork was instantly established: 1) CPT Kresefski would identify the NVA target location, and report it to LTC Hendrix; 2) LTC Hendrix told me where our howitzers should place direct fire; 3) When I received the word the howitzer(s) were ready to fire, I reported this to LTC Hendrix, and he would tell me to stand by; 3) LTC Hendrix would tell CPT Kresefski the howitzer was ready to fire, and to warn his Soldiers in the perimeter bunkers to take cover; 4) As soon as CPT Kresefski reported "All Ready" to LTC Hendrix, he gave the command to "Fire" which I relayed to Specialist 5th Class Ronald K. Haas, FDC Section Chief, who in turn sent it to the howitzer, who fired it! We fired three beehive rounds in quick succession, and these rounds broke the enemy assault. Beehive rounds were an innovation of the Vietnam War, using hundreds of one-inch long flechettes that looked like nails with four fins to help stabilize them in flight. When fired close-in at the enemy, at the proper fuse setting, the flechettes were propelled at lightning speed from the shell canister into the exposed enemy. They were very effective! A short time later, as the NVA fire resumed, LTC Hendrix ordered B Battery to fire high explosive (HE) rounds in direct fire into suspected enemy attack positions to the northwest and southwest. There was a large stand of mahogany hardwood trees, probably 25 to 30 meters from the edge of the perimeter, which provided an excellent backstop to cause the HE rounds to detonate with devastating effects on the attacking enemy. As a result, by 9:30 PM, the enemy fire had slackened considerably, although sporadic mortar and small arms fire continued throughout the night.

LT "Tappy" McEwen earned a Bronze Star with "V" device for Valor for his bravery in leaving the safety of the FDC bunker to assist in controlling the direct fire of the howitzers. Tappy relates in his Vietnam scrapbook about FSB 25: "We were under sporadic mortar fire for 28 consecutive days. On 31 January 1968 the NVA placed 500 rounds within a 100 x 200 meter area." The NVA were determined to blast us off FSB 25. They failed!

Next morning, 6:30 AM, 1 Feb 68, enemy movement was again detected northwest of the fire base, and mortar fire from 3rd/12th Inf was placed into the area. B Battery fired three rounds of direct fire at the movement, causing it to stop. An NVA soldier, attempting to fire a machine gun from a position 25 meters from the perimeter, was killed by friendly small arms fire. Two airstrikes dropped napalm within 30 meters of the perimeter bunkers. An estimated NVA battalion had attacked FSB 25, had successfully breached the protective wires, had succeeded in reaching a point only five meters from the bunker line before their attack was broken. That morning I had a chance to walk around the battery area. There were two dead NVA soldiers at the northwest edge of the perimeter, just a few feet from the parapet of one of our howitzers. They both had been carrying bamboo satchel charges, no doubt intent on blowing up our howitzers. The satchel charges were medium size rectangular boxes, perhaps a foot and a half square, and four inches deep. They were attached to a 10-foot bamboo pole. The end of the pole held by the soldier had a "pull trigger" mechanism which the soldier would pull on once he had the basket on the target (our howitzers!). Fortunately, our brave Soldiers did their job and stopped them! Amazingly, there were no US KIA that night, but 11 US WIA, none from B Battery.

After the failure of the enemy attack on FSB 25, no further ground contact occurred in the area during the period 1-10 Feb 68, although mortar and recoilless rifle fire were received almost daily.

During this period FSB 25 received 146 60m and 82mm mortar rounds, and 12 75mm recoilless rifle rounds. Twenty casualties resulted from this enemy fire, none of them from B Battery." The 6th/29th Arty Daily Log for 2 Feb 68 states "At 1245 a Div Arty H23 Observation Helicopter with WO Dyer as pilot and LT Skoglund as AO received automatic weapons fire on a low level visual reconnaissance just north of Dak To. The pilot was apparently hit with the small arms fire and lost control of the aircraft. The ship crashed just north of the hook pad. The pilot was killed and LT Skoglund received burns about the upper body. Additional injuries were not determined at the time. Equipment loss included the aircraft, one PRC-25, 2 weapons (M-16 and CAR 15), 1 flak jacket, pilot's SOI destroyed in fire."

From 11 Feb 67 to the end of the month, FSB 25 received incoming 82mm mortar rounds virtually every other day, and 75mm recoilless rifle one of those days. Undaunted, B Battery continued to shoot fire missions for the 3rd/12th Inf, and other 1st Bde Battalions when requested. With the regular incoming rounds coming in so often, we were beginning to feel a bit of a siege mentality, feeling a little numb to the "whosh" of mortar rounds passing nearby. Remember what we all learned pretty quickly in Vietnam: If you hear the rounds pass by, you're good—they've already landed! Fortunately, B Battery suffered no casualties from this series of attacks.

HILL 1338---FIRE SUPPORT BASE 5

CPT Robert Barrett, B Battery Commander, taking azimuth readings and getting map spots on NVA 122 mm Rocket positions firing on Dak To from Rocket Ridge, 13 March 68. Photo source unknown.

B Battery Howitzer firing from Hill 1338 on NVA 122 mm Rocket sites on "Rocket Ridge" that have just fired rockets on Dak To below. Photo source unknown.

B Battery rounds landing on NVA 122 mm Rocket sites adjacent to Hill 1338. SP4 John Nicolson photo.

1LT Howard "Tappy" McEwen (L), 1LT Ravin "Ray" Carpenter playing badminton on Hill 1338. Note 3rd / 12th Infantry 81 mm Mortar pit just behind. Source of photo unknown.

B Battery Command Group on Hill 1338. L-R, 1LT Howard "Tappy" McEwen, Fire Direction Officer, First Sergeant James O'Halloran, CPT Robert "Bob" Barrett, Battery Commander, 1LT Ravin "Ray" Carpenter, Executive Officer. Source of photo unknown.

Cannoneers from "The Brute" gun section on Hill 1338. L-R: PFC Louie Campbell, PFC Sam Noe, SGT John Crowe, Section Chief, SP4 Charles Lilley, SGT Nick Gillen, Gunner, SP4 Salvador Sanchez, SGT Charles Ulibarri. SGT Nick Gillen photo.

SP4 Salvador Sanchez toting his M-60 MG urging his fellow cannoneers to "Follow Me"! SGT Nick Gillen photo.

B Battery 105 mm howitzer section firing mission from Hill 1338. B Battery Commander, CPT Bob Barrett, lower left corner, looks on approvingly. Source of photo unknown.

View of Dak To Air Strip looking north from Hill 1338. Aiming Circle Rt corner. Source of photo unknown.

Well developed B Battery howitzer parapets on Hill 1338. Friendly competition between Gun Sections "Ball Buster" in foreground and "The Brute". Note 292 Radio Antennas in background. Photo Source unknown.

PFC Sam Noe and SGT Nick Gillen giving the NVA the "Italian Salute"! SGT Nick Gillen photo.

Special Forces NCO's, E-7 (L) and First Sergeant (center), names unknown, with Sam Noe visiting Hill 1338 to receive Artillery Gunnery and 105 howitzer training from B Battery. Nearby Special Forces Camp had US 105 howitzer from French Army when they were used during French Indo-China War. SGT Nick Gillen photo.

Huey Helicopter approaching LZ on Hill 1338 making resupply run. Source of photo unknown.

After the Ben Het Offensive the end of Feb 68, the 3rd/12th Inf moved to "Rocket Ridge" to halt increased enemy activity in that area. B Battery went with them, being airlifted from FSB 25 to Hill 1338, FSB 5, on 8 Mar 68. Hill 1338 was the scene of intense fighting for B Battery and 3/12th Inf during the Battle of Dak To. We were glad to be leaving FSB 25! The move to Hill 1338 went smoothly, and we were lucky to be occupying a position vacated by another unit, so settling in was relatively easy. Also, unlike FSB 25, we did not receive incoming fire from a welcoming party! The Infantry employed a heavy patrolling program (company and platoon sized patrols) to deny the enemy launching sites for his 122mm rockets. On 13 Mar 68, the morning quiet was jolted by a rocket attack launched on the Dak To Base from about 3 Km due west from our fire base on 1338. It was a clear day and we could see the rocket trails heading into Dak To and had a good idea where the launch site was located. We plotted the location in our FDC and cleared the location for firing with the 3rd/12th TOC and informed 6th/29th TOC as well. We got permission to fire. It was a fun mission to fire. We had no unit or Forward Observer near the launch area, but no problem. Since we had such good visibility from our high position on 1338, we treated the mission like a direct fire mission. We did our own Forward Observation from the firing battery, and we plastered the suspected launch site pretty well. The B Battery cannoneers got a big kick out of it because they could actually see where the rounds they were firing were landing!

The next morning at first light the Infantry made a sweep to the launch site; two sites were discovered with heavy use, and three armed 122mm rockets were captured. I like to think that B Battery's "Turkey Shoot" discouraged the NVA from firing those three armed rockets!
With the exception of our "Rocket Shoot", B Battery's first month on 1338 was relatively quiet, a welcome respite from FSB 25. Requests for fire from the 3rd/12th Inf were sparse. B Battery had a chance to "rest and refit". The 3rd/12th was doing the same. After a month of this relative quiet, I switched job positions with Captain Richard Wright. He assumed command of B Battery, and I assumed his battalion staff position of S-4. My time with B Battery was memorable!

ATTACK ON FIRE SUPPORT BASE 29

Fire Support Base 29, west of Dak To near Cambodia and Laos Border and Special Forces Camp at Ben Het. Gun Section "The Brute", L-R SGT John Crowe, Chief of Section, SGT Nick Gillen, Gunner, Unknown Soldier, SP4 Charles Lilley, Unknown Soldier, PFC Albert Lucero, PFC Orandy. Note steel pots and flak vest being worn to protect against constant NVA artillery and recoilless rifle fire from Cambodia. SGT Nick Gillen photo.

SGT Nick Gillen on Fire Base 29 just a few days before getting on a plane to rotate home! SGT Nick Gillen photo.

The following excerpts are taken from the Operational Report and Lessons Learned, 6th Bn, 29th Artillery, 30 April 1968—31 June 1968, LTC Albert Akers, Commanding: Starting on 24 May 1968, a general build-up of intelligence information began to appear in the area west of Ben Het. Much of that intelligence was first provided by the AN/TPS 25 radar situated on FSB 6 which provided seven personnel sightings on the evening of the 23rd and early morning 24th May and seven more on the 25th of May, 1968. On the 25th, the NVA initiated a program of heavy mortar, B-40 rockets, recoilless rifle, and artillery fire on fire support bases 1, 6, 12, 15, 25, 28, and 29. In one 24 hour period from 251200 to 261200 May, the AN/MPQ-4 radar emplaced on FSB 15 acquired the grid location of 51 enemy weapons sites of various types to include all those weapons previously mentioned plus some 37mm anti-aircraft weapons. At this time FSB 28 and 29 were manned by Infantry units, and in the case of FSB 29, a small contingent of 23 members of B Battery, 6th/29th Artillery, who comprised an advance party for preparation of an Artillery position, led by 1LT Ravin Carpenter, Acting Commander of B Battery. That Artillery position was being readied to counter a reported tank threat developing along the Cambodian border. If the target of the attack was as reported, the overrunning of the Special Forces Camp at Ben Het, and the main axis of the attack was to follow Highway 512, it became obvious that FSB 29 would have to be neutralized by the NVA if their plan was to stand any chance of success. By 2200 hours, it became obvious that the main attack was directed against FSB 29. Movement had been detected all around the base and sporadic small arms fire received from as early as 1930 hours. Probes were made on the west and north sides with the main attack coming up the east slope at a point where the least amount of U.S. automatic weapons could cover the area and where the artillery suffered disadvantage as far as safety requirements on the gun-target line was concerned. Despite the difficulties encountered, two FO's from the Battalion positioned themselves so that all four cardinal directions were under their combined observation and the pre-planned and adjusted fires of seven artillery batteries formed a ring of steel around FSB 29. At one point a group of NVA penetrated the perimeter wire and occupied a group of bunkers on the east side of the fire base. COFRAM ("Firecracker") rounds fired on that area for the rest of the night and was highly successful in denying the NVA entry into the breach. That group which had occupied the bunkers were destroyed and the attack was completely broken and the enemy routed by 260730 May. The 23-man advance party from B Battery, 6th/29th Artillery was formed and employed as Infantry by 1LT Ravin Carpenter, the acting B Battery Commander. Two members of that group were KIA by B-40 rocket fire while fighting from an Infantry bunker on the vital eastern approach. They were PFC Nilon K. Bacon and PFC James D. Hodges. The 1st Battalion, 8th Infantry After Action Report states that there were nine WIA in the B/6th/29th Advance Party. At our Reunion in Las Vegas, October 2012, SSG Thomas Rucker, a member of Advance Party, described to me the severity of the NVA attack. In a recent conversation with 1LT Carpenter, he cannot recall the names of the other 19 members of B Battery's Advance Party after all these years. I recently asked LTC (Ret) Richard Wright, B Battery Commander at the time of the battle, if he could recall any additional names of members of the Advance Party. He could not, but he emailed me a copy of 6th/29th Battalion Newsletter dated 15 June 1968, Number 7, entitled Howitzer Happenings. Its only article was headlined "Redlegs Become Infantry". It gave a good overview description of the battle on Fire Support Base 29. The author was not identified, but named in the article was SSG Oliver, B Battery Chief of Firing Battery, as a member of the B Battery Advance Party. So now we have identified five members of the B Battery Advance Party, all of whom earned the Presidential Unit Citation. We will continue our search to locate the remaining 18 members to ensure they, or their family members, know that they earned the Presidential Unit Citation for "Extraordinary Heroism" on Fire Support Base 29.

General Orders HEADQUARTERS
DEPARTMENT OF THE ARMY
No. 75 Washington, DC, *2 December 1969*

THE PRESIDENTIAL UNIT CITATION (ARMY). **1.** Award of the Presidential Unit Citation (Army) by the President of the United States of America to the following units of the Armed Forces of the United States is Confirmed in accordance with paragraph 194, AR 672-5-1. The text of the Citation, signed by President Richard Nixon, 21 October 1969, reads as follows:

By virtue of the authority vested in me as President of the United States And as Commander in Chief of the Armed Forces of the United States, I have Today awarded

THE PRESIDENTIAL UNIT CITATION (ARMY)
FOR EXTRAORDINARY HEROISM
TO
COMPANIES A AND C, 1ST BATALLION, 8TH INFANTRY
AND THE FOLLOWING ATTACHED UNITS:
MORTAR PLATOON, COMPANY E, 1ST BATTALION,
8TH INFANTRY
ADVANCE PARTY, 3RD BATTALION, 12TH INFANTRY
AND ADVANCE PARTY, BATTERY B, 6TH BATTALION,
29TH ARTILLERY,
4RH INFANTRY DIVISION, UNITED STATES ARMY

Companies A and C, 1st Battalion, 8th Infantry and attached units distinguished themselves by extraordinary heroism on 25 and 26 May 1968 while engaged in military operations against a numerically superior hostile force at Firebase 29 west of Dak To, Republic of Vietnam. The men of this task force demonstrated unfaltering courage and the highest degree of professional skill in successfully repelling a massive combined artillery and ground attack by an estimated regimental-size North Vietnamese force, composed of elements of the 101st and 95th Regiments with the 120th Sapper Battalion attached. At 1730 hours, 25 May 1968, the firebase came under intense enemy artillery, mortar and rocket fire. The mortar platoon and the 106mm recoilless rifle crew immediately initiated counterfire. At 2000 hours, enemy sapper squads, despite heavy casualties, managed to destroy a section of the barbed wire perimeter. Once the defensive wire was breached and several key defensive positions were destroyed, North Vietnamese elements launched a three-pronged attack and swarmed into the perimeter, managing to take control of six friendly bunkers despite heavy resistance from Company C forces and members of two advance parties. Moving quickly to the aid of her sister unit, Company A sent a counterattack element to the embattled hilltop. The combined forces managed to halt the hostile attack. Even though vastly outnumbered, the task force fought bitterly all night in close quarters for control of the vital hilltop. As close-in artillery and airstrikes ravaged enemy forces that massed on the slopes outside the perimeter, the defenders of the firebase repulsed repeated enemy attempts to overrun their positions. Throughout the battle, the enemy rained hundreds of rounds of B-40 rockets, 75mm recoilless rifle and 100mm assault gun fire on friendly positions. With the coming of daylight, the enemy was forced to withdraw while suffering heavy losses from friendly fire. At dawn the task force launched a powerful counterattack which swept the hill of all remaining resistance and restored the perimeter. Through their fortitude, superb determination, and extreme devotion to duty, Companies A and C, 1st Battalion, 8th Infantry and attached units rendered a once-powerful North Vietnamese force ineffective. Despite the numerical superiority of the enemy and the loss of key defensive positions and many personnel early in the battle, the defenders of the firebase performed their duty in the highest traditions of the military service. Their gallantry and extraordinary heroism reflected great credit on themselves, their units and the United States Army.

LIFE AT BASE CAMP IN CAMP ENARI

Base Camp compound for 6th Battalion 29th Artillery at Camp Enari, Pleiku Vietnam. Headquarters first building on left. Transit Barracks buildings lined up on both sides. Supply, Maintenance, and Motor Pool at rear of photo and not visible. CPT Bob Barrett photo.

To this point focus on the history has been on the tactical combat actions of the forward TOC of the Battalion Headquarters controlling the fires of three firing batteries, A, B, and C supporting the 1st Brigade Infantry Battalions. The forward TOC was the home of the Battalion Commander, the Battalion Sergeant Major, the Battalion Operations Officer (S-3) and his staff, and the Battalion Intelligence Officer and his staff. Nearby was the Battalion Reconnaissance & Survey Officer and his section of Soldiers (Steve Stark tells their story in his history), and the AN/MPQ-4 Radar Section commanded by WO2 Owens. These were the fighting elements of the Battalion, but there is no way they could take the fight to the enemy without support from "Battalion Main" at the Base Camp in Camp Enari at Dragon Mountain near Pleiku. I became acutely aware of this when Captain Dick Wright and I switched duty positions in April 1968: Dick from Battalion Logistics Officer (S-4) to Command of B Battery—me from B Battery to Battalion S-4. Before I begin my description of life and duties in Base Camp, I will not refer to it as "The Rear". There was no Rear Area in Vietnam, no front lines. Our Battalion units were scattered out and separated by up to 100 miles. All of our units, regardless of size or location, were subject to ground attacks, ambushes, and artillery and mortar fire at any time.

I'll begin with a brief description of the Battalion area at Camp Enari. By the time we of the second wave of replacements arrived in early September 1967, the buildings for all of the 4th Division units had been built. We had a horseshoe building arrangement, with the Battalion Headquarters at the bottom of the horseshoe, and several single level, and narrow, barracks extending from the bottom of the horseshoe down a gently sloping hill. At the bottom of the hill was a shower facility rigged in a manner to provide hot showers daily. The Battalion mess hall was next to Battalion

Headquarters. One of the barracks type buildings was configured to house the Battalion S-4 Operation. In a separate building beyond the barracks complex was a larger building which housed the Battalion Maintenance Operation.

The Battalion Headquarters Building had an office for the Battalion Commander, which he seldom occupied because he spent most of his time in Dak To at the Battalion Forward TOC. The Battalion Executive Officer, Major Carl Olson, had an office next to the Battalion Commander. He spent all of his time in Camp Enari, providing oversight and leadership to all of the activities of the Battalion happening there. Thus he was the representative of the Battalion Commander in the daily contacts with 4th Division Artillery (DivArty) Headquarters nearby, providing a conduit for information between DivArty and the Battalion Commander at Dak To.

When I became the Battalion S-4 and moved to Camp Enari, our Battalion Personnel Officer was Captain Allan Johnson. His office was in the Battalion Headquarters. He was responsible for handling all personnel actions in the Battalion, a huge task given the constant personnel turmoil in Vietnam. This included in-processing incoming personnel, out-processing departing personnel, handling awards and decorations, promotions and disciplinary actions. I noted that this was a fulltime job for him. He also was responsible for the distribution, and dispatch, of personal mail. The Battalion Mail Clerk, Specialist Fifth-Class John Tempel, had his Mail Room in the Battalion Headquarters. More about his duties soon.

I soon became aware that the Headquarters and Service Battery First Sergeant, Earnest Ferrante, was the key Senior NCO at the Base Camp. His office was also in the Battalion Headquarters. He became my First Sergeant when I assumed Command of HSB, so more about this superb NCO when I discuss that move.

When I first became Battalion S-4, I was a bit nervous because I had no previous experience with supply and logistics other than "if you need something, go ask your Supply Sergeant for it!" I needn't have worried. WO2 Harry D. Canada was the Supply Warrant Officer, and he quickly took me under his wing. Very early on he said to me, "Captain Barrett, you take the supply requests from the Battalion Commander or Executive Officer, and I'll take care of rest". Mr. Canada had years of experience in supply, and he knew how to make the system work. He also knew that if you did things by the Army Regulation "Book", especially in Vietnam, you would not get what you needed, and certainly not when you needed it. Mr. Canada knew instinctively what our Battalion needed to sustain operations in Vietnam. So about 2-3 times a week, Mr. Canada would come to me and say "Sir, let's jump in the jeep and drive around". We would drive around Camp Enari to multiple unit supply areas, Mr. Canada looking for items he needed for our Battalion. I soon learned that Mr. Canada knew all of his fellow Supply Warrants on Camp Enari. We would stop, Mr. Canada would go talk with his counterpart, share a cup of coffee, and then we would move on to the next unit. Eventually we would circle back to Battalion. Then Mr. Canada would tell me, "We need to dispatch a deuce-and-a-half truck to such and such unit, pick up the item needed, and drop off the item requested by the other Supply Warrant". Call it good old fashioned "Horse Trading". It was not regulation, but it worked and it provided critical items that would not be readily available! To set the record straight, Mr. Canada and the Battalion S-4 team used the standard U.S. Army supply system procedures, but these alone could not satisfy the supply requirements to sustain the Battalion in Vietnam. Mr. Canada's singular goal was to provide our

Battalion whatever it needed, and when it needed it! One of the items that was in constant demand was helicopter sling equipment. Sling equipment was essential to support the unit air mobile movements and resupply operations that were always ongoing in the Highlands. Sling equipment would wear out, get lost, and was subject to damage or destruction from mortar and rocket attacks. In Vietnam there was a neat way to account for destroyed or lost equipment. It was called "Combat Loss". Virtually any item of equipment that was lost, destroyed, or damaged beyond repair caused by enemy action could be written off as a "Combat Loss". There was a nifty form called a DA 444 where these affected items were listed with a brief explanation as to how and when they were lost. I helped prepare that form as Battalion S-4. Once complete I handed it to the Battalion Executive Officer for his signature, and we were done. All items on the DA 444 were no longer accountable as Battalion property, and we were now authorized to requisition replacement items. This streamlined process was a godsend and much more efficient than the bureaucratic "peacetime" process.

After about two months as Battalion S-4, I assumed Command of Headquarters and Service Battery in June 1968. Now my office moved to the Battalion Headquarters, just an earshot away from Sergeant Earnest Ferrante, now my First Sergeant, or "Top". He was the consummate First Sergeant, a WWII and Korean War Veteran. Tall and white-haired, he commanded respect from his Soldiers and Officers as well by the "no nonsense" and professional manner he carried out his duties. "Top" was the senior NCO at Camp Enari with Sergeant Major Taylor staying permanently at the forward Battalion TOC at Dak To. Regardless of Battery assignment, whenever an Enlisted Soldier was at Base Camp in Camp Enari, First Sergeant Ferrante took responsibility for them. "Top" held two formations of Soldiers daily, morning and noon, accounting for all Soldiers, and clearly stating what he expected from them at each formation. "Top" Ferrante was usually the first senior NCO a newly arrived Soldier saw when he signed into the Battalion, and the last he saw when he departed. I had complete confidence in "Top" Ferrante, he was a pleasure to work with, and he assisted me immeasurably discharging my duties as HSB Commander. "Top" Ferrante had a favorite expression when we conversed. We didn't "chit-chat" a whole lot—we both had duties to take care of. But at the end of our talks "Top" would always say "No question or doubt about it, Sir". I'll never forget it, and I took it to mean "We're on the same sheet of music"!

For Soldiers in combat in remote, dangerous battlefields far from home, probably nothing boosts morale more than getting mail from home. "Mail Call" was always awaited in great anticipation. Getting the mail out to our Soldiers was a top priority in our Battalion. After a recent reunion in Nashville, TN I spoke with Specialist Fifth-Class John Tempel who had served as our Battalion Mail Clerk. John was assigned that duty of January 1968, and his Mail Room was located in the Battalion Headquarters at Camp Enari. I asked John how he discharged his important duty of getting incoming mail distributed out to all recipients, and getting their mail picked up and sent home. No small task. Here is his story:

I had three other Soldiers working for me at any one time. They were PFC Richard Snigar, E-5 Barry Larkin, and SSG King Simmons. Between 0400-0500 every day we would depart with two jeeps to Division Artillery Headquarters nearby at Camp Enari to pick up the mail for our Battalion. Then we would bring it back to our Battalion and sort it by Battery, and put it in mail bags for them. The mail was distributed every day by any means possible! We would take the bags destined to Dak To to the Camp Enari Helipad/Airstrip and put it on a Huey, H-23, Chinook, Caribou, or

C-130—whatever was available. Often there was a truck ground convoy scheduled to depart Camp Enari for DakTo with Battalion trucks carrying other supplies as well. When the mailbags were returned to John and his team, they contained letters and packages being mailed home. The postage was "Free", so all our Soldiers had to do was write "Free" in the place where stamps normally went, and it would be on its way home!

Specialist Tempel also had a Top Secret Clearance, which enabled him to perform an important additional duty. He would sign for classified documents which arrived in the commo section, and ensure that they were securely received by the intended recipient.

One of my duties as HSB Commander was to oversee our Battalion Maintenance Section. We had many ¼ ton jeeps, ¾ ton trucks, and 2 ½ ton trucks in the Battalion. They were in constant use, traveling on very poor roads. Our 2 ½ trucks participated in the long road convoys between Camp Enari and Dak To. Our Battalion Maintenance Section did a superb job, often working around the clock to keep our vehicles rolling and performing their mission. As I experienced when I was Battalion S-4, I had in Battalion Maintenance two professionally competent leaders that got the job done. CW2 Norbert Brauner was the Battalion Maintenance Officer who ensured our maintenance effort was accomplished. When I first met him after taking command of HSB, he looked me in the eye and said "Captain Barrett, you won't have to worry about Battalion Maintenance. Our Battalion Motor Sergeant, Sergeant First Class (Frederick) McNamara, and I have been doing this a long time, and we've got it covered!" He was right!

Camp Enari was not left unprotected by artillery support. Each of the Battalion Firing Batteries, A, B, and C, provided one of its 105mm howitzers to help form a provisional firing battery at Camp Enari to provide defensive protective fire support for the Camp. As HSB Commander I was not in the chain of command for the oversight of this provisional battery. I have to assume it was controlled by 4th Division Artillery, who would have provided the preplanned target data and the Fire Direction Center capabilities. Message here is, artillery is never in reserve, and ready to provide fire support wherever and whenever needed!

PRIVATE FIRST CLASS MARINACCI AWARD OF DISTIGUISHED SERVICE CROSS

On 14 Aug 68 Bravo Battery of the 6th Bn, 29th Arty was the 105mm artillery support unit for Alpha Company 3rd Bn/12th Inf. Bravo Battery provided Alpha Company with the Forward Observer (FO) team consisting of one Artillery officer and 2 enlisted men. The Forward Observer, 1LT Daniel Neuberger, was newly arrived and recently assumed his FO duties in the Dak To area. On 14 August Private First Class Jack L. Marinacci, an enlisted Artilleryman who had been drafted into military service, volunteered to be the Radio Telephone Operator (RTO) on the FO team that accompanied Alpha Company's recon mission who relieved another recon unit. This recon unit had been in direct contact with an NVA Artillery Battalion. PFC Marinacci volunteered for the mission because he "was tired of humping ammo" at the fire support base. Within an hour of relieving the other recon unit, Alpha Company recon platoon came under heavy enemy artillery fire. The Infantry officer and 1LT Neuburger were killed in the initial attack. No other officers were in the recon unit. Taking the initiative, PFC Marinacci, who had been trained as a "cannon cocker", not a Forward Observer, secured an undamaged radio and began calling in defensive artillery fire. Throughout a long night PFC Marinacci rallied the surviving members of the infantry platoon and kept the NVA at bay with artillery barrages. The next day PFC Marinacci was able to lead the surviving soldiers to safety, and even though suffering from a head wound himself, refused evacuation until all of the other wounded were safely evacuated. Fifteen members of the recon mission were killed during this battle. Most if not all of the survivors reportedly suffer from Post-Traumatic Stress Disorder PTSD today.

For his extraordinary heroism under enemy fire, PFC Marinacci was awarded the Distinguished Service Cross, 2nd only to the Medal of Honor. He is the highest decorated member of the approximately 2,500 men who served in the 6th Bn, 29th Arty in the 4 years it was in Vietnam. PFC Marinacci has a steel plate in his head from the wounds he suffered on 14 Aug 68. The citation for his Distinguished Service Cross follows:

Headquarters, US Army, Vietnam,
General Orders No. 5679
(December 10, 1968)
Action Date: 14-Aug-68
Service: Army
Rank: Private First Class
Company: Battery B
Battalion: 6th Battalion
Regiment: 29th Artillery
Division: 4th Infantry Division

The President of the United States takes pleasure in presenting the Distinguished Service Cross to Jack L. Marinacci (US56588320), Private First Class, U.S. Army, for extraordinary heroism in connection with military operations involving conflict with an armed hostile force in the Republic of Vietnam, while serving with Battery B, 6th Battalion, 29th Artillery, 4th Infantry Division Artillery, 4th Infantry Division. Private First Class Marinacci distinguished himself by exceptionally valorous actions on 14 August 1968 while serving with an artillery forward observer team accompanying a two-platoon reconnaissance patrol in the Central Highlands northwest of Dak To. While preparing a night location, the unit came under heavy enemy recoilless rifle, mortar and small arms fire. Realizing that immediate artillery support was required, Private Marinacci ran through the murderous fusillade to the forward observation post to assist in calling in artillery fire. Racing toward the post he was wounded, however he continued to the position where he found the company commander and the forward observer had been killed in action and the radios damaged. Ignoring his wounds, he fearlessly exposed himself to the aggressor's fire until he located a working radio. Private Marinacci then assumed command of the surviving elements and made contact with the battalion commander. After leading his comrades out of the killing zone, he returned to the exposed hill top with a small group to evacuate the wounded. He left the area only after he conducted a complete check to ensure that all wounded personnel had been evacuated. Private Marinacci then led the group to another night location where he established a tight defensive perimeter and adjusted artillery fire to provide security during the night. At daybreak he supervised the extraction of the unit by helicopter. Private First Class Marinacci's extraordinary heroism and devotion to duty were in keeping with the highest traditions of the military service and reflect great credit upon himself, his unit, and the United States Army.

Recollections of Captain Richard H Wright, Battery Commander, B Battery 6 Bn 29 FA.
PFC Marinacci's exploits as cited above:

We needed an RTO to complete the FO team for Alpha Company. He volunteered to go as the RTO. If I remember correctly, he said he was tired of "humping ammo." The Battalion Commander he was in contact with was the Battalion Commander, 3rd Bn 12th Infantry, LTC Jamie Hendrix, call sign "Grizzly". He and PFC Marinacci were in direct radio contact the entire night. The Artillery LNO came from the Battalion TOC to the battery FDC and told the FDO, LT Howard "Tappy" McEwen not to interrupt any of the transmissions between them, but to take the calls for fires from Marinacci and complete the fire missions. The conversation was on the loud speaker so everyone in the FDC could hear the fire requests. Grizzly kept the conversations going in calm even tone. He knew we were monitoring the frequency, so he let Marinacci know we were in a three-way loop. For his part Marinacci also talked in a calm voice as he moved targets around his perimeter. It was a long night. As the Battery Commander, I was in and out of the FDC all night. The FDO and FDC soldiers were doing their job so I stayed out of the way.
PFC Marinacci also received the Purple Heart. We did not know that night he had been wounded. In my opinion, he should have received the Medal of Honor. Because of his Artillery MOS he could not receive the Combat Infantryman Badge either.

OUR FORWARD OBSERVERS

1LT Joseph (Joe) Knowlton with unnamed FO Party Soldier in Dak To area. 1LT Joe Knowlton photo.

1LT William (Bill) Davis leading his FO Party in the jungle near Dak To. 1LT Bill Davis photo.

1LT Bill Davis relaxing from strenuous Forward Observer duties. Somewhere in vicinity of Dak To. 1LT Bill Davis photo.

No history of our time in Vietnam would be complete without personal accounts from our Forward Observers (FO's) and their enlisted team members. The FO's were the direct link between the Infantry soldiers we supported and the firing batteries. The FO's, assisted by their Recon Sergeants and Radio Telephone Operators (RTO's) were the skilled Redlegs who called for and adjusted artillery fire to support the Infantry Company Commanders in the conduct of their scheme of maneuver, and protect their soldiers when attacked by the enemy. To accomplish this, the FO teams had to live with the Infantry, enduring the same dangers and hardships as their Infantry brothers. To put it in Vietnam lingo, our FO's had to "hump in the boonies with the Grunts"! Two of our Forward Observers, 1LT Pat McNulty, and 1LT Dave Nixon have agreed to share their personal experiences as an FO during the fighting around Dak To in their own words. We'll begin with LT McNulty's story.

1LT PAT MCNULTY, FORWARD OBSERVER

1LT Patrick McNulty, Forward Observer supporting 3rd Bn/8th Inf and 3rd Bn/12th Inf during Battle for Dak To and Tet Offensive. Pat earned the Silver Star in Nov 67 for his Gallantry In Action in the Battle for Hill 724. Pat was wounded four times during his Vietnam tour, earning the Purple Heart each time. 1LT Pat McNulty photo.

The Forward Observer (FO) Selection Process:

The requirements: (1) a pulse, (2) being of junior rank of the selector, and (3) the candidate must be a self-thinking man, capable (through their own thought process) of dropping to the ground when an explosion goes off nearby, or when bullets are fired in their direction.

Duration of Assignment:

There was always a shortage of FO's, back in 1967-68, in the Dak To area. That was an area right by the Tri-Border of Laos, Cambodia, and South Vietnam. Typically, you served until you were wounded. I replaced 1LT Larry ("shot in the back of the neck") Skoglund. Larry was Delta Company, 3rd Battalion, 8th Infantry's first FO. Delta Company was formed about a month before

I replaced Larry. The Delta CO, Captain Terence M Bell, Ranger Qualified officer had an excellent policy. You never chambered a round in your rifle, unless:
1-You were walking point.
2-You could "feel" eyes on you, e.g. you emerge from a forest, into a stream with a clearing and had to cross an open area, maybe 100 yards, before re-entering the forest. In a situation like that, when you crossed the "kill area" one or two at a time, while on the run, and you could "feel eyes", you could chamber a round, until you re-entered the forest.
3-You were being shot at.
It seems like a long-winded explanation, about nothing. Au contrair! Not all of the units I served with followed the same policy.
Another rule of CPT Bell, was WE always made/cut our own paths (not all units did the same):
Only dead men walk trails.
Walking point was a tiring job physically and mentally. You were maybe 20 yards in front of the company, you were the "early warning" group. You were walking through virgin forest, with many "wait-a-minute" vines. The men behind you, would whack away at the vines, with machetes. Your range of vision was typically less than 5 meters. So, every hour or so, you'd get replaced, and rotate to the rear. At one point, someone rotating off point didn't "clear" (remove the live round) in their rifle. The inevitable happened. That man tripped on a vine, his weapon discharged, the bullet hit a tree, and ricocheted into Lt Skoglund's neck. A Med Evac chopper was called. While Lt Skoglund was waiting for the Med Evac, Dale Young (Larry's RTO) said to Larry: "Sir I don't know if I can ask you, or not...but since you're going in, could I have your canned peaches?" So, I, 1LT Pat McNulty, became Larry's replacement to Delta company.
About 35 years later, at a 3rd Bn 8th Inf Reunion, I heard a phrase I hadn't heard in 35 years: "Shot-in-the-back-of-the-neck". Yep, it was Larry, he showed up for one of the earliest 3/8 BN Reunions. It was my first time meeting him, with much in common, we've been friends ever since.

Let's go back to: Duration of Assignment:

So, I was flown out to a fire base. We landed maybe 200 yards away from the fire base, someone came out to the LZ (Landing Zone) to escort me to the fire base. On the way, we had to crawl over a fallen tree. I was warned to take a large step to and from the tree. An NVA soldier was buried under it (not too deep) and the ground was squishy. I had arrived at the war.
The fire base was protected by one or more Infantry Companies. The fire base would contain an Artillery Battery with five 105mm howitzers (105mm batteries were organized with six howitzers, but in the 4th Division one howitzer and crew was assigned to a provisional battery at Camp Enari to provide fire support to defend base camp). A firing battery had about the same number of cannon cockers as riflemen in the Infantry companies. Each "tube" was parapitted with sandbags as were the ammo bunkers and the gun crew bunkers, with overhead protection (logs and sandbags). For the Infantry Company performing perimeter at the fire base, this was Rest and Recreation (R&R). The Infantry would increase clearing the fields of fire and run quadrant sweeps. If there were two Infantry Companies on the perimeter (not the norm), typically every other day one company would go out of the perimeter two or three kilometers (clicks) and search a quadrant of a circle, looking for any signs of the enemy making preparations to over-run the fire base.
Typically, I'd go out with a Company and we would stay out for two to three weeks. Just climbing mountains and hills, searching to make contact with the enemy, but hoping that we wouldn't. After a couple of weeks of observing nature we'd return to the fire base. Observing nature could be as simple as kneeling on one knee, while looking at your map, with your flashlight (with red filter on), and unexpectedly observing a long snake slithering over top of your boot, while it was out

hunting. The company(ies) guarding the fire base would switch off with the company(ies) coming in, and they'd get an R&R break.

That wasn't so for the FO team. We'd come in with one company and the next morning move out with the fresh unit. We were continually in the field. The Infantry MOS (Military Occupation Specialty) people were eligible for the immediately recognizable Combat Infantryman's Badge (CIB), but we Artillerymen were not eligible even though we shared the same dangers and hardships as the Infantry we were supporting. Little technicalities, like the wrong MOS, that was the biggest difference, in NOT getting the CIB. Heck, we were more Infantry than the Infantry, as we were continually in the field. The only solace for the FO team was that we were an integral part of an elite unit. The ARMY even had a special technical term reserved for our unit--"bait". When we moved in for the overnight stay, the fire base was overcrowded! There was "no room at the INN"! So, we'd move inside of, or around the 105 howitzer parapets, and sleep. The only fly in the ointment was that in the middle of the night the Artillery would fire off a few rounds of H&I fire. H&I (Harassment & Interdiction) was just firing at a pre-planned set of chosen points in hopes of hitting an unsuspecting enemy location. Without any warning, and surely to their glee, the Artillery would fire the H&I's, and interrupt your peaceful sleep!

Equipment:

Basically you'd start off every three days carrying almost your own weight:

Weight (Pounds)	Equipment Description
6	Rifle
6	Min. of (3) one qt Canteens (1 pint=1pound)
8	2 ea. 2 qt. Collapsible Canteens
3	Entrenching tool
5	Web Gear (pistol belt, harness, & bayonet)
6	Large Alice pack & Aluminum frame
4	Poncho & Liner (light blanket)
6	3 Hand grenades
6	2 Smoke grenades
15	Equivalent of about 9 C-rations (meals)
40	2000 rounds of rifle ammo (there are no ammo stores at ambush sites)
10	20 each twenty round magazines (naturally they were carried almost full-about 17 or 18 rds/mag. The magazine springs weren't trusted to feed RELIABLY under full compression tension for extended periods of time
4	Helmet (Steel pot) w/liner
7	Uniform & boots
2	Personal gear: cigarettes, razor, writing paper, mirror (shaving), & beer can opener

128 pounds total

RTO's (Radio Telephone Operators) didn't carry ammo like I did, but with their basic pack they had a 35 pound radio to carry and an extra battery about 6 pounds. (My recon Sgt and I would each carry a spare battery.) Medics had their own medical bag plus their rucksack to carry. All-in-all things evened out, everyone carried about their own weight. As resupply day grew near you were down about 12 pounds of water and an additional 12 pounds of food—life was good, just a walk in the sunshine!

What's this about everyone carried about their own weight? Most Yanks weigh more than 130 pounds! I started off in early SEP 1967 at 175 pounds. Two months later, on 9 NOV 67 I got wounded twice, one by a 60mm mortar round, then a little later by a hand grenade. The grenade caused pneumothorax (collapsed lung, not really collapsed, as they were full of blood). I spent the next couple of weeks in the hospital, in-country, and 4 weeks in Japan. After being fattened up for seven weeks, I was up to 135 pounds, and the 28" waist jungle fatigues were loose.

Being part of the Infantry Company Commanders group in the boonies has its benefits. We had extra C-rations, so we could choose among extra meals. Typically, it meant we got a carton of the cigarettes we liked! Some stuff just didn't appeal: white bread, scrambled eggs, etc. No problem, we'd puncture those tins of crap with the P-38 can opener for the insects and toss them.

A BEER can opener! Yes, we'd get a cold 6 pack every night—in your dreams! Reality, only once did I see ice in the field. the 1st SGT brought out a duffel bag with ice and soda, everyone got ONE, that was it. I missed soda so much that I vowed, never again. Every night since I returned from Vietnam I take a soda up to my nightstand.

We'd use the can opener to fascinate ourselves with fabricating a furnace, to put C4 (white plastic explosive) in. We'd burn it to heat our C rations.

Some units weren't as closely supervised as they should have been. Those idiots actually took C4 out of their defensive perimeter Claymore mines (to heat their C-rats). Many deaths were attributed to this practice when a battlefield analysis was carried out. Rob the claymore of explosive, and instead of the ball bearings being blasted out at 1500 feet/sec, they only move at 100 feet/second. Death of Americans is the result—actions have consequences. We'd get C4 blocks for blowing trees, so supplies could be lowered to us through the jungle canopy. We had our personal supply, maybe equal to about three packs of cigarettes, for ration heating. Others carried a brick or two of C4 for tree blowing. If you carried C4, you were NOT to carry the blasting caps. Well, I heard about a guy in the unit (I was with another unit at the time) who was carrying two bricks of C4 and blasting caps. When they were on the move, an NVA popped out from behind a tree with a flame thrower, and they both got each other. The NVA got a full magazine and the GI got lit up. Three GI's ran to him, he was on fire and they were trying to get him untangled from his rucksack and web gear, thrashing about. One of those helping got so badly burned that his leg had to be amputated. The kicker: the guy risked his life, lost a leg, a never even got a medal! All due to someone violating the rules. Stupid things were continuing to extract their toll—usually death!

Another night we had a double decker! It was frowned on to take a crap inside your night perimeter. The correct way was to do our business as soon as the night position was selected during daylight and you were digging your foxhole and clearing fields of fire. While near the end of fields of fire being cleared, then do it. Well, someone waited until almost dusk, then went out, and coming back in he was taken for a gook and shot dead. Later that night the FO with the other company started to call in fire in response to listening post reporting activity. I immediately called in CHECK FIRE on the other FO. Had this FO gotten the rounds he wanted we would have tree bursts going off overhead. He hadn't checked the angle of shells passing overhead from the location of the battery he was requesting fire from. Another Artillery Battery from another fire base could have accommodated him. Fortunately, my Company Commander, CPT Bell, backed me up and my CHECK FIRE was upheld. Next morning an Artillery Major flew out, shot an azimuth with a compass, took off, and that was the end of it. Everything you do or fail to do has consequences. I'll bring up later the consequences of a tree burst.

Stupidity is not inherited. Young men in battle die before passing on that gene. We were blowing trees for a clearing area that supplies could be lowered, via chopper, down to us. When the last "Fire in the Hole" warning was issued, I raised up from behind a log to photograph the event. How I survived, I do not know. Gigantic pieces of wood and splinters the size of railroad spikes were flying at me. I almost exited the gene pool at that point!

Terrain:

It varied all over the place. One time we were in an area with boulders all over the place, 20 to more than 30 feet in diameter, just strewn all over the place. The same area also had something looking like chimneys. We took a break, I was a "newby", and I plopped down on my rear against one of these pillars with my rucksack. About half way through my cigarette, I engraved in my mind what a fire ant chimney was, to the amusement of all the on-lookers. Other times, tall grasslands (six+ feet)). Then dense bamboo. Defoliated areas. Only high growth timber areas with almost zilch in low growth foliage. B-52 areas, cratered, and just smelling of death. Then, the traditional triple canopy jungle. I have some pictures with maybe a 15-foot view, five men in the picture and you're lucky to spot three of them.

Poor Charlie, he came South, probably pushing a bicycle full of supplies, knowing he would be away for years, no R&R to look forward to, and forced to fight against B-52's, Puff the Magic Dragon, Air & Artillery superiority, and no air re-supply. He did a fabulous job, under very adverse conditions. NVA regulars is all we ever came upon, no villages, no Viet Cong. Had they had aircraft and real artillery support, not just mortars, it would have been a different outcome. Anecdotal, they got hyped up on dope before doing Banzai type attacks-- you'd be crazy not to! They had been in the area for over 20 years. We'd spent the day going up Chu Goungot Mountain, and at the top we find a stairway six feet wide with logs staked as steps ALL the way down the other side WITH HAND RAILS! Another time, we found a vertical shaft, about six feet square. We started digging, and after six feet found a row of logs flat to the ground, and six feet below that another row of logs perpendicular to the first, then another six feet..., yet another six feet…, there was no end in sight. Finally we decided to blow it with C4. Five or 10 minutes later a plume of smoke/dust was seen rising above the trees about a quarter mile away. We moved out, never bothered finding out what is was all about.

As stated before, we were a new outfit, replacing the original boat people (INF that came over by boat in '66 vs airplane in '67). I arrived about a month after the transition, so I know nothing about them sharing/not sharing lessons learned. It wasn't until our first big battle of Dak To that we learned the NVA were big on snipers tied in trees. We didn't see any palm trees around, like the Japs tied themselves to in WWII. We were in triple canopy jungle! It was an expensive lesson! Being in triple canopy jungle is approximately like bamboo, or other scrubs clogging your vision from ground to 10 feet high, then young trees sprouting out 10 to 40 feet high, followed by mature growth reaching 100 or more feet high.

Try this analogy. If you're on the shore looking out onto the ocean, in a fog, you may not see a boat out there. Conversely, if you're on that boat, you may be able to see through the fog to the people on shore. Height makes a difference! Trooping through the jungle, you usually see nothing. However, put yourself up in a tree, and you can easily see through the lowest canopy.

Miscellaneous Experiences and Observations:

When you first enter Vietnam at Cam Rahn Bay you get our name tapes and patches sewn on. Kiss that money goodbye. When you hit the boonies, you start with a nice uniform, Ranger Tab, rank, and crossed cannons. Three weeks later you get a re-supply of DX (Direct Exchange) clothing. None of your uniforms with patches are anywhere to be found, you root through the clothing for anything that's not too small, and that's your outfit for the next few weeks.

About every three days, along with a re-supply chopper, mail catches up with you. You read your letters and then burn them. You weren't supposed to keep them. They could be used as propaganda, should you get killed or captured. At night, sometimes we'd pick up their propaganda station playing music, and then directly speaking to our unit: like "Welcome back CPL Jenkins from D 3/8, how was your R&R in Australia?"
At night you'd hear the fu*k you lizards. They'd make their call in unison, and it sounded exactly like "fu*k you".

In some areas you'd be in a forest facing out on one side overlooking a sea of grass fields. Charlie was out there, you could see them moving about using flashlights. I'd borrow CPT Bell's starlight scope and observe them clearly. I was an expert shot. In college where I was in Reserve Officer Training Corps (ROTC), they had an Army rifle range with fancy .22 caliber rifles. Every day I'd shoot 50 or 100 rounds between classes. At ROTC six-week summer camp training, I came in as 2nd best shot out of about a thousand. There was no way to mount the starlight scope to the rifle. I asked CPT Bell to get me an M-1 D rifle. It was a sniper version of the M-1 rifle, so I could shoot at them at night. Fortunately that never happened, the rifle was longer & heavier, not good attributes, in the jungle. And bullets weighed about three to four times that of the .223 we fired from the M-16. We moved out of that terrain, and it would have just become an albatross about my neck.
Being in the boonies was almost a Biblical experience. Sometimes, Charlie would make little probing attacks. It was Jericho, they would blow horns and bugles. Something that was so primitive, that it was unnerving. It worked, hearing dozens of the bugles going off, all around you. Mealtime was about the highlight of the day. Tabasco sauce was used freely, everyone carried a bottle of it. Boned chicken, Spicy Beef, Ham & Limas (my favorite from Ranger school, as it came in a large tin, most people didn't like it, too fatty). Consequently, I could easily trade my Beans & Franks for Ham & Limas). The other meats, save Beans & Franks, were in containers only half that size. But soon there was no novelty to wear off, you had immediately eliminated 1/3 of the C-rats as unappealing (scrambled eggs) and eating the other 2/3rds became a chore. You had them, with and without chopper dust; with and without ants and bugs; with and without Tabasco sauce; hot and cold. You became very picky, eating less and less. That contributed to weight loss. A few (very few) times we got a few MRE (Meals Ready to Eat), ah ambrosia! Better known in Vietnam as Long Range Reconnaissance Patrol (LRRP) Rations. Add a little water to

the spaghetti pack, re-seal it and place it inside your shirt to heat up. I can't recall a single mess hall hot meal served in the field, unless possibly when we came into a fire base.

When I started off as an FO, I had an air mattress, forget the fact that it weighed about 5 pounds! I had a little duct tape, but eventually realized I woke up daily with zero air. That was the end of that. Sleeping on bamboo stubs, rocks, roots, etc. was a challenge. Doing it in the rain just added more variations of miserableness. And absolutely, we had no pillows. The jungle was very hot, but at night it became very cold in the Highlands. One poncho liner didn't cut it. My Company Commander, CPT Bell, had an Army sweater. I was envious, so were others. I remember one morning it was so cold, I was shivering. We were due for re-supply that day. All I had left was a partial canteen of Kool-Aid. I tried making coffee out of it, but threw the mess away, it just wasn't drinkable.

I remember another morning waking up, we had moved into a recently bombed area, with deep bomb craters. We had been re-supplied with food and ammo the previous evening. We had some contact the day before. We had a little ceremony for a few men killed and I guess we were given a mini R&R and allowed to stay an extra day in the area. It was a very cold morning, and someone started a fire in the crater with the C-ration cardboard, and then tossed in the empty ammo crates. The warmth felt good, but for some reason I moved away. As I moved away an explosion went off. Someone tossed in some rubbish/crate that had a grenade in it, and it cooked off, and several got wounded. Everything has consequences.

Early in the RVN war, out of folklore, some Infantry LT's wouldn't let their men go out on patrol wearing after shave lotion, or smoking cigarettes. It wasn't in touch with reality! Recall, we wore the same clothes for three weeks at a time. When our guys went on patrol and made contact, they reported they could smell the enemy coming, and quickly set up a hasty ambush. I'm sure smelling the enemy worked both ways. It wasn't the shaving lotion smelled, it was ten guys that hadn't had a shower in a month, wearing the same clothes for three weeks. Secondly, as iterated earlier, your gear weighed about your own weight and no one carried an ounce more than they had to carry. Just stationary paper, flints and lighter fluid for your lighter. Unless you count the tail end of the monsoon season, except for one stream bath, I can't recall a shower in a two-month period.

Leaches:

How they penetrated bloused trousers, I do not know. And you didn't have to cross streams to get them. Once you had dug in for the evening, it was entertainment time. Today, what method will I use to remove leaches: burn them off with a cigarette; use my bottle of 95% DEET on them; or just plain pull them off?

Moonscape:

I won a free travel package, including free helicopter rides. It was unbelievable the similarity (save the green color) between the moon and the craters in South Vietnam. We dropped more ordinance there than we dropped in all of WWII, and it showed. The teak forests couldn't be harvested for generations, for wood products, due to the shrapnel.

Sixth Sense:

Either you develop it, or you increase your chance of premature death. A couple of weeks of mountain humping ended, we pulled into a fire base, about mid-afternoon. The fire base had attracted an entrepreneur sniper. Being an invincible Ranger (I still had a lot to learn about supposed invincibility), I loaded up on about a dozen grenades, had them hanging from my pockets, took about 20 spare magazines and went out with a half dozen others. We never found the sniper but I definitely came real close. I know I had his eyes on me, the only reason I wasn't shot was because he would have killed himself with all my grenades. Several other times, when moving outside the perimeter, firing Defensive Fire Concentrations (DEFCONS), I could feel eyes on me and my FO party, so we quickly moved back into the perimeter. We all felt those eyes.

Officers:

A real Bell Curve!

As I mentioned earlier, you spend about three weeks out in the boonies, then return to a fire base. You drop off the company you were with and leave the next morning with the fresh company that had been guarding the fire base. Consequently, I got to work with all of the companies. CPT Falcone and CPT Taylor were both West Pointers, I believe. One had been the best man for the other's wedding. They both had been to Ranger School, I believe, and knew their stuff. CPT Falcone was an extremely rare individual. He was really loved and admired by his troops. He was one of less than five individuals that I have met in my life that was so well liked and competent that you would do anything for. He in turn would do anything possible for you. Neither would guess they would both be dead in a month. CPT Bell was another Ranger graduate, competent and well respected.

Then we had another CPT who wasn't fit to lead a pack of Cub Scouts. I wound up being his FO, once. We were out for a two or three-week period. During that time we missed a re-supply (food and water) for some reason. So we had to go the next day, until evening, with what we had, close to zilch. I believe I had a breakfast like in Ranger school: three or four packets of salt and pepper. About midday, point radioed back they had come across a stream. Just about everyone dropped their pack AND RIFLE! They went forward to fill their canteens. They filled their canteens and started drinking immediately. That's right, no Halazone tablets. Halazone tablets were to purify the water, you'd drop in one or two and wait 20 minutes before drinking.

I moved forward to the CPT, and started yelling at him, told him to get security out, men on the far side of the stream, and up and downstream, and get the rest of his men back to their weapons and gear. I told him in a very loud voice that he wasn't fit to lead a pack of Cub Scouts. I thought I was going to get court martialed over that incident! He finally got security posted. Security upstream reported finding two dead NVA in the stream! When that word reached the troops it was almost comical to watch their reaction. They were still drinking water without dissolving Halazone tablets, but now they started swallowing the tablets whole!

When I was first assigned to Delta Company, 3rd of the 8th INF, I realized the enormity of the task I was assigned to. My first fire mission was scary, just like anything you do in life, speaking into a radio, talking to a microphone, driving a car, or calling in fire. In this case it was calling in DEFCONS at a known point, calling in Artillery about the four quadrants, 300 meters out from your nighttime position. I had to write out my call for fire, the first time, so as not to screw it up.

As mentioned earlier, I carried about 2,000 rounds of 5.56mm ammo, weight about 40 pounds. The coincidence wasn't lost on me that a SINGLE 105mm projectile, weighing 37 ½ pounds, was equal to the firepower that a SINGLE ground pounder carried.

Infantry is known as the Queen of Battle, and Artillery is known as the King of Battle. When you got into a fire fight, or had an objective to take, or were the centerpiece of a numerically superior enemy, artillery will cause the majority of casualties. Simply put, an FO can put out in 10 to 15 minutes as much firepower as an infantry company can put out in TWO DAYS. If the FO lives and is competent and can bring it in close without killing his own men, the Infantry Company lives. I often adjusted fire to within 25 meters of our troops. Firing into the woods does nothing, firing onto the enemy is what counts! The Infantry Company Commander and the Arty FO have enormous responsibility for the lives of the entire company.

I was determined to be the best FO out there. One mandatory element of that task was to be an excellent map reader. I ALWAYS knew exactly where I was, on the ground. I was within 15 feet of where I was on the map, even when we were using maps provided by the Army that had the caveat that the map was an interpretation/projected guess as to what things were like on the ground. Army cartography was excellent!

When stuff hit the fan, the FO could replicate the firepower of an entire 80-man infantry company exponentially. An 80-man company was carrying the firepower of 80 105mm rounds. When an FO is firing five howitzer tubes in "fire for effect" at the enemy target, those five round volleys can quickly be shifted and re-fired any number of times. In about 15 minutes the FO can replicate the entire firepower of an Infantry Company.

Charlie was tenacious, his objective was us. He "hugged" us, if he was within 25 meters of us he had a chance of living because that was the safety limit for "danger close" Artillery fire for our troops. If he was 50 meters from us, he was toast. Fortunately, the artillery I was firing was done very professionally, they leveled their bubbles between every round, and the fire came in accurately. Never did firing Artillery within 25 meters of my position cause any casualties.

Having the Infantry fire mortars was a different story! We were coming in to a fire base one night, but didn't have enough daylight to make it to the fire base before darkness. We were close enough to have Artillery support and Infantry mortars. After registering Artillery and Infantry DEFCONS (Defensive Concentrations), all seemed well. About six hours later, the listening posts reported enemy sounds. Listening posts were made up of a group of about three men, stationed about 30 meters outside the main body position--they are your early warning line, your eyes and ears, at all four quadrants. They each knew how far in front of their position I had adjusted their DEFCON. I got reports from them telling me there was activity about 50 meters in front of them. I'd adjust fire, in 50-meter steps, towards where they wanted it. The challenge for me was how do you keep track of exactly where the last round came in at each of four quadrants when you have contact at all four quadrants simultaneously (Four quadrants refer to 360 degrees covering the entire company perimeter, divided into four parts). This was never taught in the Field Artillery Basic Artillery Course. I had to quickly come up with a method. Fortunately, the method came to me very quickly: Get out your notepad, put the main group in the center, and mark each outpost, from the center. Then as you fire for that outpost, mark the last round fired with a higher number or letter-- problem solved. Well, not quite. I had fired an 81mm mortar concentration, and the enemy was being driven right towards it. Perfect. I called for 5 rounds at that concentration. What I got was horrible. One round landed fine, the next three landed inside our perimeter, and the last round fell outside, below our position. They DIDN'T level their bubbles between each round! Somehow, we took no casualties. The mortar squad claimed they fired for speed, they thought I wanted it fast, so they didn't bother leveling the bubbles. I had no faith in the mortars from then on. I only used them once more, months later, when I felt we were being almost over run. Even then, I called for check fires on the mortars to determine if the rounds landing nearby were from our mortars, or the enemy's!

Hill 724, BATTLE OF DAK TO

Early NOV 1967, we were told by higher-ups they knew about where the enemy was. Army brass called upon our specialized unit ("Bait") to locate them, only 2 or 3 Regiments. Since we're talking only a few enemy regiments, we won't need more than A and D Companies, 3/8 Inf. So we do a helicopter combat assault (CA) into the tri-border area, lucky us, and scared sh*tlesss! Such great odds!

4 NOV-- we CA (Combat Assault) in. Dig in night positions, cut down trees, and put in overhead cover, clear fields of fire.

5 NOV-- A Company sends out a patrol, a firefight erupts and a LT in A Company gets killed. Then CPT Taylor gets killed trying to recover the LT's body. The gook that killed the LT was captured. LT's from A & D Company, including me, determined that the SKS rifle captured from the gook should be sent back to his widow so she could hang it over her fireplace. It seemed perfectly reasonable to us AT THAT TIME. Combat does affect one's thinking!

6 NOV-- A and D Companies tear down their bunkers and overhead cover. Yours truly, leaves behind a "trophy" for the NVA, in the bunker dirt. A Company, takes the lead in moving off our hilltop towards the objective, Hill 724. As D Company follows, A Company reports finding booby traps, trip wires with hand grenades. Then, an Artillery round makes an air burst over D Company's position. I immediately call in a "Check Fire". I wasn't firing, but it ceased all firing in the area. Was it someone firing H&I outside their Area of Responsibility? I still don't know, other than it was "Friendly Fire". Result: Sergeant Greenwood of D Company was killed, Walter Gross lost his leg. The advance stops, and we pull back to our caved in positions. We re-dig our positions, and on one of the digs with my entrenching tool I hit my left behind trophy (make a note to clean the shovel well).

7 NOV-- A and D Companies tear down their bunkers and overhead cover again. More reports of booby traps, the FO party is in about the midpoint of the two companies, Alpha in front, Delta, now several hundred meters off of last night's position. We are about five meters to a muddy stream, before starting our ascent, and all hell breaks loose. A .51 caliber heavy machine gun opens up on us. We can't see where the machine gun is. No idea as how far up the hill the gun is located in dense triple canopy jungle. The .51 is terrifying, and everyone automatically has dropped to the ground. Some of us lucky ones have a one-inch diameter piece of bamboo to take

cover behind, others a fallen log, others nothing. Now we are being hit by 60 mm mortars. We are not aware, but there were snipers in the trees. We didn't know it, but the NVA Regiments were acting on two goals: (1) find and utterly destroy an American unit, and we were targeted to be that unit, (for Public Relations story, in the newspaper) and (2) to move out of Laos and Cambodia and down to the coast for the TET offensive.

This day had all the warning signs that we were going to have a very bad day. I had called in fire on the NVA before. I had no problem with trying to kill them, and I never killed anyone that didn't need killing. My job was to protect the men I served with. But usually it was a one-sided event or just a fortuitous event, for our side. Once I was on Hill 724 I could actually see out over the trees and into a valley. We were within sight of the tri-border area, and in a Free Fire Zone (FFZ- only NVA and us in the area). I was firing my first DEFCON of the night. A smoke round was always fired first, as a precaution to make sure you were exactly where you thought you were, on the map. If that round was OK, you followed up with HE (High Explosive). Well, when the smoke round popped, someone in the valley popped a green smoke grenade. I quickly checked with CPT Bell that no friendly units were in our area of operations (AO). He confirmed it. Then I verified with the Fire Direction Center (FDC) and they also confirmed no friendlies, so the area got plastered all over with HE. Fortuitous for us, bad for Charlie. In this incident, Charlie popped a smoke grenade to confuse and confound us. He was out to buy time to di-di (leave) out of the area, but unfortunately for him, I got two confirmations in about a minute. The artillery pumped out a half dozen "battery ones". Each battery one was a volley of five 37.5 pound projectiles of HE being unleashed, about 190 pounds of explosive. That was repeated, by "walking it around" moving the impact area 50 meters a half dozen times. The artillery accomplished this in just a couple of minutes. So in two minutes 1140 pounds of HE went off in an area roughly 200 x 200 meters. That's about equal to 400 hand grenades going off in the same area. Sheer terror for the recipients. We are now on the receiving end of NVA mortar fire. I had been wounded by their 60mm mortars, nothing serious, unless you're a Timex watch. The fragments blew my Timex watch into my wrist, and in this case the Timex "took a licking, but didn't keep on ticking".

As soon as the .51 caliber machine gun opened up, and I found my 1-inch piece of bamboo to take cover behind, I was calling in fire on the .51. I was scared, we were in heavy action. Probably CPT Bell; each of the platoon leaders, Platoon Sgts, squad leaders, and everyone below them had that sickening feeling in their stomach. I know the .51 opening up on us was the root cause, but I think the feeling in your stomach is caused by the "flight or fight syndrome" which causes the blood to drain from the extremities to your trunk body, so you don't bleed out fast. When the mortars started coming in on us just a few minutes later, the fear compounded. They had us between a rock and a hard place. I had to deduce where could they be firing from. It could be from anywhere. They could have made clearings to fire the 60's from anywhere they chose. In that case, the rounds could have come from anywhere. So, I had unlimited points on the map to fire at-- all I needed was unlimited ammunition-- plus all the time in the world to silence them. By the same token, the NVA most likely never made any clearings, because they never expected us in their backyard. When we landed in the area a couple of days ago, if they had felled trees to make clearings for the mortars to fire from we would have heard the trees falling. So I eliminated the millions of possible places. Instead, I figured that Charlie would always move into an area vacated by the enemy if they knew the enemy (us) was in the area. Why would they do that?--to evaluate how competent we were. The NVA would move into our previous nighttime position and observe what they saw. If they noted that we merely slept on the ground they knew we were an easy mark. Conversely, if they observed that we dug foxholes, we earned more respect from them. Further, if we dug bunkers, felled trees, put in overhead protection against grenades and mortars, and cleared fields of fire, i.e. removed grass, saplings, and brush, NVA respect for us further increased. In Delta Company, with CPT Bell, we did the full scope whenever possible. It wasn't always possible, as some areas were devoid of trees, so no overhead cover, for example.

Still on 7 NOV We didn't know it, but Dak To was to become the largest, bloodiest battle of the war to date. The first night at the knoll, at the base of 724, Alpha and Delta Companies were counterattacked. Alpha Company got the worst of it and was partially overrun. During the night a resupply of ammo was dropped. It was in a large cargo net. As fate would have it, it landed in the wrong spot. Can't blame the chopper pilot. It came at night, during the counterattack, and there was no LZ, just triple canopy with hundred-foot high trees. The wounded were together at a collection point, and the cargo net full of ammo landed on top of them. The next morning the cargo net was used to haul out twenty-one of our dead.

During the ensuing three and one-half days, the remnants of Alpha Company were evacuated and Bravo and Charlie Companies were brought in. CPT Terry Bell was the single surviving officer left of the twenty original officers. I believe we had 57 KIA's in the 3/8 Inf Bn. For wounded numbers you can normally project four wounded in action (WIA) to one killed in action (KIA). That means there were approximately 300 casualties out of a 500-man battalion in four serious days of fighting.

I was medically evacuated (Med Evac Helicopter) from Hill 724, arriving at the 71st Field Evacuation Hospital in Pleiku. As I laid on a gurney, I was greeted by a Doctor who said he was Lt Col Major! I'm thinking, what kind of new rank is that? He was a Doctor with the rank of Lt Col, with the last name of Major, very confusing at that point. Next, the Doctor had me sit up, took a scalpel, and without further ado, cut a three-quarter inch long cut in my back. Then he pushed in a plastic hose, for about a foot. Then the hose is pushed forward until it breaks through the chest wall membrane. It hurts! An empty Cider bottle (they didn't have plastic bottles back in 1967) is attached to the hose which acts as a drain for the blood in my lungs. It took my breath away when the Medics yanked it out about two to three weeks later. Between the 71st Evac and Cam Rahn Bay Hospital I spend about two weeks, then off to Japan Army Hospital at Tachikawa AFB, Kisini Barracks for a month of recuperation.

For his Gallantry in Action during the battle for Hill 724, LT McNulty earned the Silver Star. His Citation reads as follows:

DEPARTMENT OF THE ARMY
HEADQUARTERS 4TH INFANTRY DIVISION
APO San Francisco 96262

GENERAL ORDERS
NUMBER 8204

28 December 1969

AWARD OF THE SILVER STAR

TC 329. The following AWARD is announced:
McNULTY, PATRICK J. JR. FIRST LIEUTENANT FIELD ARTILLERY United States Army, Btry C, 6th Bn, 29th Arty, 4th Inf Div, APO 96265

Awarded: Silver Star
Date action: 7 November 1967
Theater: Republic of Vietnam

Reason: For gallantry in action while engaged in military operations against an armed hostile force in the Republic of Vietnam. First Lieutenant Mc Nulty distinguished himself while serving as an Artillery Forward observer with Battery C, 6th Battalion, 29th Artillery, 4th Infantry division. On 7 November 1967, Lieutenant McNulty was attached to Company D, 3d Battalion, 8th Infantry when the unit came under intense enemy attack near Dak To. Sensing the incoming mortar rounds were coming from two adjacent hills, Lieutenant McNulty immediately called for fire on these positions, at the same time organizing and directing the friendly elements in the area. Moving around the perimeter, Lieutenant McNulty was severely wounded by a mortar round. Fighting off unconsciousness, Lieutenant McNulty continued to direct accurate fire. At one point in the battle the enemy tried to overrun Lieutenant McNulty's position and while engaging the enemy at extremely close range, Lieutenant McNulty was again wounded. As the enemy began to retreat, Lieutenant McNulty began organizing litter bearers and assisted the medics in administering aid, though seriously wounded himself. First Lieutenant McNulty's extreme courage, outstanding performance, and exemplary devotion to duty are in keeping with the highest traditions of the military service and reflect great credit upon himself, his unit and the United States Army.

Authority: By direction of the President, under the provisions of the Act Of Congress, approved 9 July 1918.

FOR THE COMMANDER:

OFFICIAL: JOSEPH E. McCARTHY
Colonel, GS
Chief of Staff
J. F. LASSITER
CPT, AGC, Asst AG

There is a "feel good" piece that must be added to this story. I'll say up front that LT McNulty was awarded the Silver Star on 28 December 1969, but he did not know he earned the award until December 2017! Our 6th Bn/29th Artillery member Research Coordinator, LTC Ray Harton, US Army Retired, is solely responsible for correcting this long standing and unacceptable oversight. Here are highlights of how Ray made things right for Pat and his family:

During his year in Vietnam during which he served the entire time as a Forward Observer, Pat was wounded four times, for which he earned four Purple Heart Medals. Pat was looking for a listing of his four Purple Heart Medals on the 6th Bn/29th Artillery website but could find no record of them. Pat called Ray knowing he was the current 6th Bn/29th Artillery Database creator and manager. Ray told Pat that the old database had been shut down, and he should use the updated one Ray had created. When Pat reviewed it, he saw the four Purple Hearts he knew he had earned, but also noticed another award listed as "SS". Pat asked what this award was, and Ray replied "that's a Silver Star"! This was the first time Pat knew he was awarded the Silver Star. Ray asked Pat if he had received his Silver Star and he responded "No, and I have no documentation to prove I earned it". Ray went into "high gear" to track down Pat's Silver Star. Through a strange twist of events Pat was listed in a newspaper section titled "Beyond the Call" as one who should have been awarded the Silver Star. Using copies of this newspaper article, Ray made a formal request through the Freedom of Information Act for a status on Pat's Silver Star. Eventually they responded and confirmed Pat had been awarded the Silver Star, backing this up with official documentation. The crowning moment of this Odyssey is when Pat was presented his Silver Star at a special ceremony in Philadelphia, PA led by Pennsylvania Senator Pat Toomey with Pat and his extended Family at the Ceremony. Pat's granddaughters sang at the Ceremony!

In closing of this "good" story, we must give tribute to LTC Ray Harton, US Army Retired. This is a classic example of Ray's devotion to his fellow "Can Do" Band of Brothers. Fifty plus years after his brave service that first year in Vietnam, leading the way for the rest of us, Ray has continued to serve his "buddies". Thanks for your Service, Ray!

On 30 December I was shipped back to Nam after recovering from my latest wounds. I got a pair on new fatigues, still have them, they can't be worn. In SEP 67 was 175 pounds and had a 36-inch waist. in DEC 67, I was 135 pounds (after being fattened up) and down to a 28-inch waist! When I arrived back in NAM, I was offered a job escorting Ann Margret and such around the country. Like a fool I insisted on returning to my 6/29 Arty-3/8 Inf unit. But 3/8th Inf has all its FO's. 3/12th Inf needs one. Guess where I go? Bravo Company, 3rd Bn 12th Inf! It was a good unit, good leadership, good men. Only problem is, they see more than their share of combat.

We are just putting in a new fire base-- FSB 25. FSB 25 is a 6th Bn 29th Arty fire base, commanded by CPT Robert Barrett, a southern gentleman. They support both 3/8 and 3/12 Inf.
Building a fire base is great duty, no mountain rucksack humping, just some interrupted sleep when they fire H&I's, during the night.

Eating C-rats has become a real problem. I can't stand any of the meats, save Ham and Limas. Gone are all the others from my preference list: Spiced Beef, Boned Chicken, Beans & Franks... all gone. Now I'm living off crackers and cheese, fruit cup, pound cake, and a few others. About every other day one of the two companies goes out a few clicks (kilometers) and makes a quadrant search- translation: move out, search one quadrant of the compass at a distance of 2 to 4 km away from the fire base. Make sure there are no signs of Charlie sneaking up on us for an attack. Out in the morning, no rucksack, back at night, maybe cut or blow trees for an alternate landing zone. Absolutely delightful, without a ruck.

We're now up to 19 JAN 68, I've been back almost 3 weeks. Last night, SP4 Timothy M. Nafe, who is my Recon Sergeant and I give each other haircuts. Tim had a pair of manual hair cutters. Today is our turn to make that quadrant sweep. Things as simple as a quadrant sweep, can be dangerous. Just a few days earlier, we did a sweep. During it we started cutting down trees to clear an alternative / backup LZ. At the time, I stopped and watched a tree falling. It wasn't falling in my direction, so why were people motioning to me? I found out. A large tree limb above me was dislodged and headed towards 'moi'. I saw stars during the daylight. Fortunately, my helmet was on my head, or it would have been split open. The Army does have concern for its people. I got a free helicopter ride to an aid station for an x-ray. No problem, so back to the Fire base.

Well, time for that little walk in the forest. No Rucksack! Hooray, again! Just your web gear, harness, canteens, ammo pouches, rifle, and an extra 20 mags of ammo, and some C-rats for a picnic lunch. Very uneventful, went out maybe 4 clicks, then point calls back to the company commander. Point spotted a couple of trail watchers and asked what they should do next.
Trail watchers were NVA regulars, just sitting at the cross point of two trails. Their job was to direct any other NVA troops as to where to go. This was their home, their backyard, they were entitled. The Company Commander and I both responded back "Shoot them". I then said to the CO, I'm calling in Artillery, been in this stuff before, and it can turn to shit real fast.

The Infantry quickly took the signal and started their picnic lunch. Bringing in DEFCONS took 5 or 10 minutes. My Recon Sgt, a young man, 20, 21 years old, just didn't understand why we were at war. I had explained the Domino Theory to him, he still didn't understand --smart man. The Domino Theory was popular at this time of the Cold War. The thought was that when countries were lost to Communist dictatorship, then the country adjacent to it would then be more susceptible to Communist takeover, followed by the next country. Thus, it was important for the US to stop the march of Communism in Vietnam.

Tim was a man who also had been wounded before. I called him over and asked if he had ever "adjusted fire" before. He said, "No", I then explained today is your lucky day. Since returning to 'Nam this was the first time I was adjusting fire. It was a good time to do some cross training. My previous FO party had observed me calling in fire, and I hadn't thought to actually have them do it. After having been wounded a couple of months ago, I realized I wasn't as immortal as I thought. So, we started our lesson.

First of all, since he was going to be doing the firing I handed him the radio mike to talk directly to the Artillery Fire Direction Center in B Battery on FSB 25. Tim had no idea of where we were on the map which was a normal expectation. I showed him, where we were and how to translate the map coordinates into code. We used code on the assumption the NVA were listening in. Uncoded grid locations could be used against us. I told him, we bring in the first round 500m from our position, just to make sure we hadn't made a mistake map reading. So what did he want to do? He said "1 HE round @ XXXXX". I explained, NO. the first round is always a smoke round. You'll never see the smoke round in this triple canopy jungle, but you'll hear it "pop" when the smoke "goes off ". Having smoke come in first, it is just a double check that your map reading is correct. The Infantry would be very upset with a HE round landing on them during adjustment!
Tim calls in Smoke. I reply, "That's perfect, remember that sound. Now what?"

Tim says "HE", I tell him correct, tell them "repeat with HE". He repeats that into the mike.
Now what? I ask. Tim says, "Drop 200", I say: "we only drop in 100-meter increments, just to be doubly sure". Tim calls in "drop 100", and that comes in fine. Now what? Tim says: "drop 100", I say "good."

He calls in "drop 100", it comes in fine. Now what, Tim says: "drop 100". I say: "No, at 300 meters, we only drop in 50-meter increments. We're at "danger close" fire @ 300m".

I explain we're about OK, but we want to make sure all five tubes are landing at the same point. To do that, we request the FDC to repeat fire with one tube (base piece), and have a second tube fire with a 2 second delay after the first, and we'll listen to where they land. Tim speaks into the phone what I said, they fire, he notices the differences in the impact. Crunch.....crunch. So what's next? Tim replies, bring the 2nd tube to left 50 meters, and add 50 meters. Good call, have them repeat fire, same sequence, with 2 second delay. Tim calls it in, the rounds come out crunch, crunch. Tell FDC that is perfect, both tubes are hitting the same point. We will stop firing, we have artillery at a known point, and will adjust fire from there. Request FDC record that point as "Target". Tim calls in, "Record as Target". I tell the Infantry Company Commander, "we're good, we've got Artillery".

Unbeknownst to us we had stopped right over top of NVA Regimental headquarters (yep, another 2,500 of them). They were 90 feet deep in the ground. When we shot their trail watchers they came topside. While we were adjusting fire, they were surrounding us and getting in the trees. Triple canopy jungle is like fog. When you're on the ground, you've got scrub stuff and vines reaching up maybe 10 feet, then saplings trying to get higher into sunlight, then trees. From the ground you can't see the sunlight, or even 15 feet high. It's like fog at the seashore. if you get higher up into the trees you can see down through the brush and saplings relatively easily. The Company Commander hears the OK to move, his men get up, and as Tim takes his first step, he takes the first bullet, right through the heart. Since Tim was doing all the talking, the NVA spotted him through the "fog" and tagged him as someone important, bringing artillery into their playground. Anyone who carried a pistol, carried a radio, carried a carbine instead of a rifle, or one stupid enough to wear officer insignia brass, was an automatic prime target. They were an officer, NCO, or someone in the command and control group. Take out the leadership and communications, and the rest become directionless, and easy to rout or overcome. Tim was dead before he hit the ground, he still had the radio. Bullets were like a swarm of hornets kicking up the ground all around me. I yelled to someone to throw me the radio. The Company Commander had informed higher up we were engaged and they were questioning him as to how serious it was. I remember as clear as seconds ago the Company Commander replying: "I'll tell you how serious it is, I'll put a dead man (Tim Nafe) on the phone to tell you!" Immediately I started adjusting the artillery concentration off to the side where the trail watchers were because the incoming heaviest fire was from there. As I'm firing the 105's, I get another call: Hello, I'm 155's-- can I be of assistance to you? It was fabulous news. Before I could reply to that offer, here comes another: I'm a 175 unit, can we help? I quickly decided to give two of the units coordinates on the map to make a straight line (like our left flank, encompassing maybe two grid squares on the map) and told them to fire anywhere along that line and add 100 to 200 meters or so depending upon their caliber, in depth. That way I could adjust an entire block of fire at a time. Great, now I have two large caliber batteries of guns making a sort of corridor for us. At time, I can concentrate on moving the 105's around with precision. Here comes another offer: I'm 8-inch, need assistance?

It took about four hours to advance to the rear. Ambush coming out of the ambush, into the next ambush...Sheer hell! I wasn't with any group, I was like 99% alone. I'm not aware of how that specifically happened, partly I was at the tail end because I didn't want to bring in fire too close to us, those big calibers have quite a lethal range. Another reason was stopping to try to keep my bearing on the map. I was alone until two GI's without weapons came up to me. They asked if they could have my rifle. I asked where were theirs? They replied they had been carrying back a wounded guy, who died. I hadn't fired mine the entire time, I was too busy with the map and radio. I asked them, what could they give me? They offered a grenade. I said "OK, but you guys swear you're going to stay with me and protect me?" "Yes", they said. I swear, within two minutes I was alone again!

We were getting close to the fire base, I had to stop firing the 105's because we were too close to the fire base. They did a great job (CPT Robert Barrett's B Battery) when I was in range. Now, I'm using mortars. The small arms fire has died off, not much of it now, and thank God nothing directed personally at me. But I have a problem with the mortars, some are dropping in our area. I call for a "Check Fire" but the mortars want to override me. I tell them to stop, I need to verify that they are not the cause of rounds in our area. They stop, but we are still taking incoming, so it is Charlie, not short rounds from our mortars. I turned our mortars back on.

As I near the safety of fire base, I want to stab my leg with my penknife. I have had enough terror, anything to get away from this. But I don't do it. Finally, I get to within 100 meters of the firebase. I see a Major, he is in a clean uniform with polished boots. He offers me a drink from his canteen, it was ice cold Scotch. Unbelievable! I don't like Scotch, don't even know if I swallowed any. All I wanted to do was get inside the fire base and collapse.

I later find out our KIA toll was seven, but the wounded was in the 35 to 40 range. We were well mauled, I think only 2 radios came back, mine and the Company Commander's. Not a single machine gun made it back, along with many, many missing weapons (maybe they jammed, maybe they were dropped to carry the wounded). Tim Nafe's body was the only one brought back. The fighting was intense, and niceties like retrieving bodies has a lower priority when taking high casualties.

I heard that each day for about a week the Infantry went out to recover the other six KIA's and were driven back day after day. This was even after B-52's came in and carpet bombed the area. Charlie finally left when he wanted. I went to the Infantry command post bunker where my rucksack was and found my rifle and ammo that I had given away had been returned. That was nice.
We naturally expected a counter attack that evening. One of our .50 caliber machine guns opened up. The gunner had spotted some gooks. The blood instantly pooling in the stomach feeling returns.

The fire base was still under construction. The area had been a nice forested hilltop. That's a lot of trees and fields of fire to clear. I believe there was a Caterpillar tractor and maybe a backhoe. Simply put, it was a herculean effort in moving trees, digging in howitzers, sandbagging and parapets. There was a big shortage of bunkers with overhead cover. In the bunker I was in there were probably over 20 men, and that was sitting on your behind, with your knees drawn up to your chest. It was too claustrophobic for me. If the attack is forthcoming, I can't do anything trapped one/third of the way back in the bunker, so I exit the bunker. There's an RTO in the front of the bunker, in touch with the Listening Post (LP) men. I tell them I'm moving to the outside edge of the bunker, to call me the second any LP reports anything. The attack didn't come that evening. The NVA dropped just three 82mm mortars to wake us up. Unfortunately, the first inking I had was the explosion at arm's length from my head. An 82mm mortar is about the equivalent of 7 hand grenades. A sixth sense warned me a split second before it exploded, and I had both eyes open. I lost an eye (eyelid OK, it was open), and almost lost the other eye from a shell fragment 1/16th inch from the eyeball. My forearm was across my forehead, otherwise it would have taken off the top of my head. Instead it took out a baseball size chunk of the arm, cut two of the three nerves to the hand/fingers and broke the arm in a few places, cut another hole on the leg by the femoral artery, and blew out an eardrum. The only other problem I had was about forty 1 to 2-inch shell fragments and 400 smaller pieces which found a home inside me. My nickname is "lucky". Anyway, I wasn't unconscious, I got up, and had a problem seeing. I didn't realize it was blood, I tripped, and had a problem getting up, not knowing I had a problem with my arm. I then got up and went to the front of the bunker and asked if there was a medic inside. Someone said "yes", so I said, "come outside, I've got some work for you!" I tell you, combat changes your perspective. I was flown to an aid station where they took out a lot of the big pieces, then flew me to the 71st Evac Hospital again. This time, I got the drop on Lt Col Major, I said hello to him and asked if he remembered me. Then I went to the OR. The next day Lt Col Major told me they had to en-ulcerate the eye because it was just like Jello. I was overjoyed, because he said the war was over for me. I can't say how many times I closed the other eye, which was almost swollen shut anyway, just to see what it was like to see with only one eye. Kind of foolish, but it's life, reality is variable.

About ten days later, I'm still at the 71st Evac, not stable enough to move. Word filters down that the South Vietnamese guards stopped and arrested one of the South Vietnamese help. She was caught pacing off distance from the flagpole to a building. That afternoon none of the South Vietnamese hospital workers reported for duty. It was just a coincidence, I guess, that the hospital was mortared, rocketed, and had a ground attack that breached and partially over-ran the hospital. I got my 4th Purple Heart because I was wounded again in this attack. My buddy Larry Skoglund, "shot-in-the-back-of-the-neck", was also there, but he had to be satisfied with only two Purple Heart awards.

For about forty years after Tim Nafe got killed, I felt responsibility. At some point, probably after visiting THE WALL in Washington, DC, I looked up the information on Tim and found out his hometown. It was about one and a half hours away from my home. My family visited his mother, I told her how her son died. It took a burden off her. Her son's funeral was closed casket, and she always figured he was captured and tortured. I assured her that was not the case. Why it was closed casket I cannot understand, one bullet to the heart, and body brought back the same day. For years we visited her close to the 19 JAN date, and went out to dinner. Finally, she passed. Then one day it hit me: Had I been on the radio that day, I would have been killed. The 3/12 Inf Company Commander was wounded that day also. In all likelihood, had I been in Tim's place and KIA, instead of B/3/12 having 35 to 40 WIA and seven KIA, it's quite possible without the fire support I provided as FO most of Bravo Company would have been killed or wounded. That finally gave me peace.

1LT DAVE NIXON, FORWARD OBSERVER

1LT Dave Nixon, B Battery Forward Observer, checking out his caliber .30 US Carbine at Fire Base 25 near Dak To. 1LT Nixon photo.

I have been in contact with 1LT Dave Nixon for several years during which time he has shared several emails describing his experience as an FO. With his permission he has allowed me to share these with us. To get the true flavor of what Dave is describing in his emails, I will stay as true as possible to his verbiage. Here it is:

"I served as one of your illustrious FO's from December '67 until I got wounded on February 26, '68. You might recall, FO's were hard to keep around. When first assigned I joined the Battery on the hill overlooking Ben Het (Fire Base 25). I probably met you a couple of times, but I stayed mostly with the companies I was assigned to. If I went and humped with C Company (3/12th Inf) for instance, when they came in, I turned right around and went out with D Company or some such. There was, I was told, a huge FO shortage! I only knew one other FO in the Battery who went to Ben Het for a 'short time' and never came back to the hill. Hence, I went out and humped his company also. I think his name might have been Pruitt. On the night of 31 January '68 (beginning of Tet Offensive) we were humping right at the base of the tri-border area. We scurried back at dawn the next day (actually we were up and moving about 2:30 AM) and headed back to Fire Base 25 due to LP (Listening Post) alerts and helped retrieve the NVA bodies from trees where they were stuck from beehive (flechsettes) penetration and, after pictures, helped bury the remains.

After the close call of being overrun at FSB 25, the Grunts (C Company) and my FO team came up with a test plan that would discourage ground attackers. Diagram by 1LT Dave Nixon.

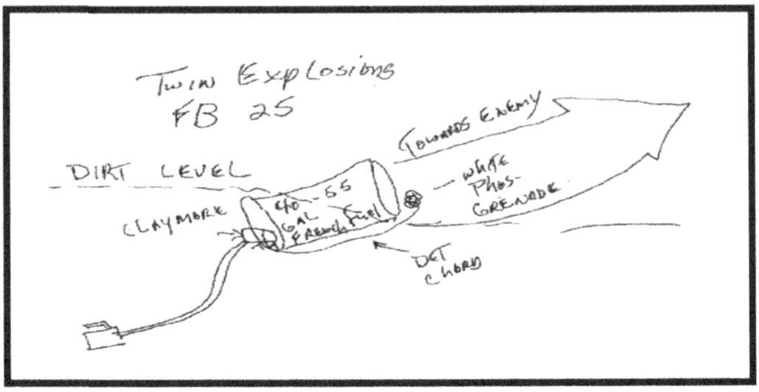

Four barrels of very volatile French fuel oil were located (too old to use in machinery reliably I guess). We measured off about 60 paces from the last gun, through the wire, and dug two trenches to hold a barrel each. Then we placed the fuel barrel in the trench, backing it up with a claymore (mine), wrapped in detonation (det) cord and wrapped the other end of det cord around a white phosphorus (WP) grenade resting just beneath the top of the front of the barrel. Please refer to the diagram. The theory was, when the claymore was detonated, it would propel the French fuel out of the exposed end, being ignited by the exploding WP grenade and send the fiery ball over the heads of the enemy, starving the air of oxygen. The test was conducted by a Grunt Platoon Sergeant. Test worked famously! I was told they got to deploy it either the next night or several nights later with a very high success rate. I don't know because I was out humping again with another company. But while I was there between the big attack and humping again, I manned the wire each night (FO eyes for the 8-inch howitzers) with several satchel charge probing attacks. We were glad to get back into the bush! You folks (on the fire base) had very, very large testicles to go through that stuff almost nightly as sitting ducks. My hat is off to you and the Battery. At least in the bush, we could shoot, move and communicate!

I had always wondered what the BC scope was for, though. And I remember how there was 1 VC/NVA who threw a mortar round into the fire base vicinity around lunchtime once in a while. Kept everyone on his toes. No rest for the weary!

On 26 February '68 we came out of FSB 25. I think my call sign might have been Pegleg Ladder 27. We were so short of FO's that I rotated through several companies, getting new call signs all the time, and happened to be in B Company, 3rd/12th, on 26 Feb 68 when I was wounded. 1LT Bill Davis was the follow-on FO in A Company (call sign 47). Bill and I flipped a coin at the extraction point to start off this new campaign and I lost. As far as I know he ended up being on the last ship out of the Extraction LZ. While in the air, although I was seated on the first ship scheduled into the LZ, the next helicopter in line landed first for whatever reason. I was on the second helicopter that got shot down into the LZ. I didn't get wounded until several of us attempted, and succeeded I might add, to secure the first NVA trench line up the hill from the downed helicopter. The helicopter had landed into a 500-pound bomb crater made LZ. All 16+ of us on the ground had our feet touching inside the bomb crater. I was fearful that a single, well placed grenade would wound us all at once, so I took my RTO (unremembered name), a couple of grunts, a door gunner and the pilot of our ship with me into the trench line, about ten meters up the hill, where we held off the enemy's attacks for over an hour. All of us were inside the trench line firing and I was calling in artillery the best I could. I called in a Danger Close, one round Delay fuse artillery fire mission to see what we could see. It turned out we were on the Gun/Target line. The round came in, hit the trees overhead about 20 yards in front of us. We three had our heads down (the door gunner had run back to the LZ). That round made a mess. So I asked for High Angle—Repeat. That put the target out of range. So I moved the round out 100 meters. It flew over the hill we were on and landed about 400 meters beyond the hilltop. The grunt (Infantry) Battalion Commander was screaming at his Artillery Liaison Officer to get rounds on the target. He never fired a round while I was there. The enemy, in a desperate attack, heaved five to seven grenades all at once towards our position. We exchanged grenades with the enemy. They had CHICOM grenades, we had our fragmentation grenades. All enemy grenades but one landed outside the trenches we were in, doing no apparent harm to anyone. One grenade bounced off my helmet and landed in the trench about six feet behind me, causing my wounds. There was collateral damage also to my RTO and the pilot. Since I was first in line in the trench I got it the worst. Before it blew I yelled to everyone "Grenade—cover your ears!" I lost both eardrums with perforations and a big hole in my backside at the beltline. Right after the grenade attack several enemy charged our position and we repelled them. With me somewhat badly wounded, along with the RTO and pilot, we decided to fall back and regroup at the LZ and hook up with the other company (A?) coming up the hill behind us. I was the last to leave, trying to hold my pants up while doing fire and maneuver! As we left the trenches I killed at least one other NVA who managed to get in behind us and was firing from a bamboo thicket. He was between us and the downed helicopter so he had to go. I picked up his SKS rifle to take with me. After reaching the downed helicopter site I was given medical aid for my wounds. When laid on the stretcher, I had the SKS lying next to me with my ruck in trail with my camera in it. I never heard what happened to my wounded RTO or the wounded pilot. I had passed out shortly after getting aid and awoke in the Dak To field hospital being hustled and chastised by the Artillery Battalion S-3. In the end he took my SOI/SSI and never said goodbye, thanks, kiss my ass, or anything. What a putz! The next time I woke up I was in Pleiku Hospital." I was alive, but my ruck with camera and the SKS were never returned to me!"

MY RECOLLECTIONS OF VIETNAM—SP4 JOHN NICOLSON

(Editor's Note: Specialist Fourth Class John C. Nicolson served in the 6th/29th Artillery from Sept '67 to Sept '68. He served bravely in the Battalion Liaison Section with the 3rd/12th Infantry. Here is the story of his year in Vietnam in his own words. Sadly, John passed away in October 2016, just prior to our Nashville reunion which he and his wife planned to attend.

SP4 John Nicolson, who served his entire year in Vietnam as a member of the 6th Battalion, 29th Artillery Liaison Section with the 3rd Battalion, 12th Infantry. SP4 John Nicolson photo

INTRODUCTION

The following is a collection of my recollection of my time in the Army. For the most part I've tried to match the events with the locations in which they occurred. I think I've done pretty well in this regard, but at this writing it's been close to forty years ago.

A lot of events may seem pretty insignificant but I figure that if I carried the memory of them this long they're worth recording. So too the names of my fellow Soldiers. My Army days were a very intense time for me, kind of like Alice in Wonderland. It began suddenly and ended the same way. In between it was a kind of insanity, especially basic training and Vietnam, and I think it was because of the friendships we forged with each other that we were able to see it through.

Most of the things I've described I was witness and/or participant to. Some I heard first or second hand from someone I would trust to tell me the truth. The sequence—I don't really remember the sequence after all this time-- but wrote the events as I figured they probably occurred.

VIETNAM

I have very little recollection of Oakland, California. I don't think I stayed there more than a day. It was plenty time to put me on KP duty, though, and the last duty I had before leaving the United States was mopping floors.

There was hardly anyone there I knew. Brian Bishop was there though, and we managed to sit together on the plane. We stopped to refuel in Honolulu and they let us off the plane for about forty-five minutes or so. One point of interest on the flight was that we flew over Midway Island, scene of the decisive battle that turned the tide against Japan in World War II. We landed at some air base near Manila in the Philippines, it was probably the Manila airport, and stayed there several hours. They had to replace an engine that had quit working on the flight. Anyway, they had a hanger set up with chairs and snack bar and we were pretty much confined to it. By the time we got to Vietnam our flight had taken more than twenty-four hours.

Vietnam was very hot and very muggy. We were put on busses with heavy wire screens over the windows to keep bombs or hand grenades from being thrown in, said the driver. They reminded me of prison busses, the whole thing very ominous. We drove in convoy, a machine gun carrying jeep in front and another in back, through the outskirts of Saigon to something called the 90th Replacement Battalion. This was a processing center that matched the new arrivals with assignments. After filling out paperwork we were shepherded to wooden barracks similar to Fort Knox except they had screen windows, no glass. I was given a top bunk with a lleak in the roof right over it. I had been wide awake since I had awakened in Oakland and was exhausted. I slept like a rock and woke up soaked.

There was nothing much to do at 90th Replacement but attend the many formations they called each day. At each formation they called out several names. These would fall out and be led away, taken to their new assignment. Brian was called out and was led away. I never knew where and never saw or heard from him again. I was called out in a late afternoon formation and taken to an airfield. We were marched up onto a C-130 Transport plane, made to sit on the floor with our knees brought up to our chests, our duffel bags between our knees. The duffel bag and knees of the man behind served as our backrest. It was dark when we took off and the plane was very noisy. We flew to Pleiku and were driven, probably in trucks, to Camp Enari. We were now members of the Fourth Infantry Division, the Ivy Division.

The Fourth Division ran a five-day course for new arrivals to familiarize us with the peculiarities of Vietnam. It was designed to get us to think for ourselves so we wouldn't just follow set procedure that was taught in the States. If ambushed don't-- don't charge the bushwhackers thinking he'll run; he won't. Shoot back from cover and call artillery in to kill him. We were given many practical ideas to on how to deal with various situations that might arise. It was set up like basic training so there was a bit of running and other physical training. This was late August, the wet season in Vietnam, and everything was covered in mud, especially us. This course included an overnight combat patrol. We had real bullets and had a real chance, though very slight, of actually running into the enemy. One month or so earlier one of these patrols had witnessed a mortar attack on some position though they didn't get involved. Anyway, I thought it was a lot of fun and a great way to end the course.

SIXTH BATTALION, TWENTY-NINTH ARTILLERY

At the end of the course I was assigned to the 6th/29th Artillery, Headquarters and Service Battery (HSB). I had always thought of the Headquarters Battery as the ones who kept the firing batteries supplied—supply clerks. I was trained to operate the guns, be a real soldier, and being assigned to a Headquarters Battery was quite a blow to my ego. It looked like I'd get home alive now though! Wouldn't be able to hold my head up when I got there.

One of the first things I had to do was to go for an interview with the battery commander. He was sitting behind a desk, I was standing at attention in front. He may or may not have welcomed me but I remember the rest of the interview. "Do you know how to operate a radio, Nicolson?" "No Sir". "Do you think you could learn?" "Yes Sir". "Good, I'm sending you forward". Those words hit me like a blow, I had quickly gotten used to the idea I would live and this sounded like a death sentence. I had heard the sound of the base camp guns firing and knew death lurked out there somewhere. I'd just been told that's where I'd be going, into harm's way. I felt sick to my stomach as I left the commander's office.

The next thing on the agenda was to draw all our gear. This was to include rifles. Those staying in base camp, there were a number of us new arrivals, would draw M-14's, those going out to the field, M-16's. I thought it should be vice-versa.

As I mentioned this was the rainy season and it rained several times per hour, very hard. You could see the rain coming, look up to see the cloud it was coming from, endure it for three or four minutes and watch it go away. We were all wet and muddy but we had a parade that afternoon after lunch. I went to the barracks to change into clean clothes. There were a number of civilian workers, Vietnamese, at one end of the building but I took little notice and changed. A couple other guys came in and asked me where I changed. I told them "Right Here". One of them pointed to the Vietnamese and said "In front of these people?" I looked up and for the first time realized they were all women. We all had a big laugh at that.
We marched in the parade, at the end we stood in formation for an address by some big wig officer. The band was playing, the rain was pouring down and some Sergeant screamed "Taaaake Seats!" We all plopped down into the mud on our butts.

Within the next day or so we new arrivals who were going "forward" were ordered to report to the helicopter pad for a flight to Jackson Hole. The helicopter flew low and fast along a stream. At one point both door gunners opened fire simultaneously. I looked at where they were shooting, one side anyway, and there didn't seem to be anything there. Then I noticed the one door gunner was grinning from ear-to-ear. This was just a little show for the new guys.

Camp Enari was the base camp for the Fourth Division. The Fourth also had a number of smaller camps that headquartered the various Brigades, two or three, Jackson Hole was one of them. Around the Brigade Headquarters were placed artillery batteries in much smaller camps called fire support bases or simply Firebases. Each Firebase served as the headquarters of an Infantry Battalion and so it was from the fire base that the Battalion's four companies were given their missions and supported by the artillery firing battery. The Firebases also served to support, directly and indirectly, the Brigade's Headquarters. At Jackson Hole we were taken to a large tent with several cots inside and given some candles to see by. A Sergeant told us where to go if we were attacked. A couple of weeks earlier Jackson Hole had been rocketed and a couple of new guys had been killed. The next day I was sent to see a Major, the highest-ranking officer I had ever spoken to so far, along with another new man who had just arrived that day. I recognized him as someone who had just gone through the little five-day training program. Anyway, the Major gave us a little speech about how hard a life those who served on the guns led and told us the scores we had on our examinations, the ones I had taken at Fort Wayne or Fort Knox, I don't remember which, qualified us to work in FDC (Fire Direction Center) or Liaison. Would we be interested? We both accepted and Ken Marshall and I would be working closely together for the next year.

That afternoon I was told I had guard duty that night. That kind of surprised me since I was new and in transit, passing through as it were. I was assigned to a fighting bunker, one with a firing slot, with two other men, none of us from the same unit. One was a combat engineer, and the other was an infantryman. Both had been in Vietnam a couple of months so I considered them to be experienced veterans. A trench ran through the bunker with an entrance-way on either side. When firing from the slot, or slit, you would actually be standing in the trench. The area in the back was used for sleeping. The one on duty would stand outside the bunker, still in the trench, so he could see and hear better.

I had drawn last shift and clearly remember laying in the back of the bunker, it was dark now, and talking to one of the other men. He asked me how old I was and I replied it depended on whether it was past midnight or not, this was my twentieth birthday. I drifted off to sleep and was suddenly awakened by my very nervous companions. It was my turn and a listening post, two or three men sent out into the wilds to listen for any approaching enemy, had heard movement, so be especially alert. They were scared so now I was scared. They sent me outside the bunker with my M-16. They handed me the switch to fire the claymore mine, a type of bomb that had been planted out in front of the bunker a short distance, and a hand grenade telling me not to shoot my gun because the flash would give away our position. Needless to say, I was scared to death now. Luckily nothing came of it, but what a way to spend a birthday!

The next day a little three-man "bubble" helicopter came to take me out the Firebase. One of the men we'd be replacing, a man named Casidy, met me at the landing pad and helped carry my gear. He took me to the TOC, Tactical Operations Center, and introduced me to Lieutenant Gorka, the Liaison Officer. There was an action going on, one of our Infantry units was in contact with the enemy, so I was told to wait outside. As I sat in front of the TOC I noticed a very distressed man coming and going out of the TOC, apparently being sent as a messenger. On one of his trips he noticed me sitting there and said "You're lucky don't have to work here." I replied "This is where I'm going to be working", wondering what I was getting into.

Lieutenant Gorka had a tent with a covered foxhole right next to it that he decided I should sleep in. That night I would be spending several hours with the Fire Direction Center. These were the guys who figured which direction the guns would point, what angle their barrels would be set, and how much powder would be put behind the projectile in order to hit the target. Lieutenant Hamblin was in charge, I was basically there to learn radio procedure. The one thing noteworthy about LT Hamblin was that he was married to a woman from Holly, Mich. After several hours I was told I could leave now. Stepping out of the FDC bunker I was in complete darkness with only a sketchy idea of where LT Gorka's tent was. Trying to find my way I walked smack into a deep hole. I climbed out and eventually found the tent. After that I was sure to memorize all my routes in a given firebase before dark!

After a couple days at FDC I was sent back to the TOC. Ken Marshall had flown out to the firebase the day after me and had been trained in radio by Casidy. The TOC consisted of the box off an Army truck about the size of my backyard shed. Across the back was a shelf with a number of radios and field telephones. To the left as you entered was another shelf with two radios and two field telephones. The radios across the back were operated by the infantry, the ones on the left were the artillery liaison's. The Infantry here was the Third of the Twelfth ($3^{rd}/12^{th}$) which operated four companies (A,B,C,&D) and were supported by B Battery $6/29^{th}$ Artillery which fielded five 105 millimeter howitzers.

Liaison's job was to keep constant track of where friendly troops were. Each Infantry company had an Artillery Forward Observer. He would call in for Artillery Support for the company as he and the Infantry Company Commander felt needed. We would check his request for safety. We wouldn't allow the first round too close to him or allow it to fly over friendly positions. In most cases B Battery would do the firing but in some it was better to have a neighboring battery do the shooting. Liaison would determine this and make the arrangements. The Liaison Officer would also survey the area by helicopter and determine where the enemy was likely to shoot his artillery at us from. He would then record and chart them (put on a map overlay) and give them to the firing battery so the FDC could work up the details for hitting them. These were called DEFCONs, short for defensive concentrations and whenever we or one of our infantry units were being hit with artillery, usually mortars, we would put the overlay on the map and decide which of the targets were safe to shoot at and which weren't. This was called "clearing the cannons". We would also plot various targets to be shot during the night while the enemy was moving around. These were called H&I fires, short for Harassment & Interdiction. Sometimes they actually hit someone. We know that because our infantry patrols would come across bodies close to shell craters. At one time we killed five enemy soldiers with one round while firing H&Is.

There were a number of infantry radio operators who worked in the TOC. They were under the overall charge of Sergeant Williams, and older man ready to retire. He liked to call himself "Mad Willie" and his radio operators he called his "Willikens". There were also a number of other Sergeants and Officers who had various duties in the TOC. I better explain that the small radio shed was just part of the TOC. It was always sandbagged into a complex bunker that contained a number of rooms.

The Battalion Commander of the $3^{rd}/12^{th}$ Infantry was Lieutenant Colonel Volmer. His second in command was Major Stiner. Major Stiner was from the South and kind of an "Old Country Boy" very similar to Andy of the "Andy Griffith Show". After the evening's meeting with all of the Officers to plan tomorrow's activities he would, in turn, have a conference with us radio operators to fill us in. Major Stiner went on to be General (4 star) and was in overall Command of the Panama Invasion.

LT Gorka's radio call sign was 40 (Four Zero), Marshsall's was 40-B (Four Zero Bravo), Bravo part signifying his rank as E-4, mine was 40-C (Four Zero Charlie), the Charlie part signifying my rank as E-3. I eventually became a Bravo but always thought of myself as 40-C. The Forward

Observers with A, B, C, and D companies were 47, 48, 49, and 50 respectively. We got to know 47, 49, and 50 pretty well, but our 48's kept getting killed or seriously injured.

We moved a couple of times by truck to set up different firebase sites. LT Gorka and the "old" liaison radio operators went home. Captain Hill took over as the Liaison Officer. Captain Hill was fairly likeable, he was new to the job so he was pretty tense for a while until he learned it, but he was a fast learner. He loved his coffee and since he was a Captain it was always made available to him. That meant it was always made available to us too so there were no complaints from me. Jack Enright came as a replacement radio operator so now the 40-department consisted of four new guys. In fact, most of the 4th ID was in the process of being replaced by new guys.

The enemy contact we had at this time were with small groups of Viet Cong who carried a variety of make-do weapons left over from other wars. They usually managed to get away. If our 'grunts' (infantrymen) killed three or four it was considered a big thing. One evening one of our FO's called in claiming he was observing many lights moving around. Enemy troops usually moved at night, if they believed it was safe to do so. Lieutenant Colonel Volmer was really excited, massed enemy troops could mean a big body count if we played our cards right. He had us call up all the artillery batteries that could hit the target. This included two or three 105 batteries, at least one 155 battery and an 8-inch battery like the one I served on at Ft. Hood, and at least two mortar sections, an 81 millimeter (small organic) and a 4.2 inch (large organic). (Note: "Organic" was Infantry "insider" term referring to their own "artillery" (mortars) "organic" to their battalion organization). They all started firing and our FO confirmed they were hitting the target and adjusted the firing to hit the whole area. He also claimed the firing had no effect on the lights; we kept shooting. LTC Volmer was thrilled, we were really kicking the hell out of them! It seemed like every 30 seconds I was hearing "Ba-Ba-Ba-Boom" as our five 105's from B Battery fired almost simultaneously, and in between was the "Bang! Bang! Bang!" of the large and small organics, mortars. No change on the lights. This activity was being repeated on at least three other fire bases. Still no effect on the lights. LTC Vollmer was jumping for joy. I was getting skeptical, if all this artillery were falling on me would I not turn off my flashlight so they wouldn't know where to shoot? Quick, decisive action is drilled into the Army's Officers, common sense was something they were supposed to bring from home. No one was going to express doubt to the Colonel. We kept firing. Eventually we stopped shooting, we may have been running low on ammo. The next day they found many blown up trees, a few dead monkey's and nothing else. The "lights" they had seen were determined to have been reflected lights from Pleiku. It was one red-faced LTC Volmer who left by helicopter that morning to explain his actions to the Brigade Commander.

DAK TO

In November, 1967 we were sent to Dak To. Dak To was a Special Forces Camp near the Laotian and Cambodian Border. It lay on a valley floor pretty much surrounded by jungle covered hills, some as much as 4000 feet high. Initially we set up like any other firebase, taking over the command bunker as our TOC. After a day or so Brigade moved in and took over, Brigade Liaison taking over the duties we would normally assume. For a few days we had little to do except guard the airstrip that ran beside the camp.

The communists had been fortifying the Dak To area with at least four regiments of NVA (North Vietnamese Army) regular soldiers, completely equipped with modern weapons and artillery of their own. During the last month or so they had fought a number of ferocious battles against our 173rd Airborne Brigade. In one of them a paratroop company was overrun by an NVA battalion with the loss of 76 dead, many of them murdered by the NVA as they lay wounded.

As the 4th ID Commander, General Peers, later wrote, "Nearly every key terrain feature was heavily fortified with elaborate bunker and trench complexes--he was prepared to stay". Yes, the NVA controlled the high ground and we were going to take it away from them.
One day we started receiving enemy mortar rounds. They weren't falling very close, it looked like we might actually be out of their range. I remember standing out in the open watching them explode way outside the camp but close to the civilians who always seemed to hang around the soldiers looking for hand-outs or to sell something. Major Stiner (Operations Officer, 3/12th Infantry) called me over to him and pointed out an area of a hill out in the distance he thought the gooks were firing from. He told me to run over to a nearby tank and direct its fire against that area. This was an M-42 Duster which fired twin 40-millimeter pom-pom guns. This was quite a thrill and honor for me and I had the Duster pepper the area pretty good. The mortar fire stopped. This was my first experience of being under enemy fire, it would hardly be my last. Soon after, possibly the next day, Dak To again came under enemy shelling. As one historian wrote "The flat crack of a 75 mm pack howitzer spat shells into two C-130 transport aircraft on the air strip. The two aircraft were destroyed by fire. Enemy fire fell on the ammo dump near the air strip and with a tremendous earthshaking roar the ammo dump began a chain reaction explosion. I had been on duty the night before. I slept through the whole thing. That was on Nov 15 ('67). To the Southwest of Dak To was Hill 1338 which had a commanding view of the camp and from which the NVA planned to establish mortar and rocket positions. On Nov 17 ('67) two companies of the 3rd/12th Infantry moved out to take the hill. They were Alpha and Charlie companies. Lieutenant Colonel Jamie Hendrix had taken over command of 3rd/12th Inf and the history books say this was his show but I distinctly recall it was LTC Volmer's voice I heard giving the orders. This being LTC Hendrix's first battle LTC Volmer probably went along as advisor. Anyway, it was a big battle against dug-in troops and required air strikes along with the usual artillery support. The hill was taken.

On Nov 20 ('67) the 173rd Airborne Brigade began their assault on Hill 875. They finally took it four days later after suffering horrendous casualties. I was across the airstrip from where they were loading the bodies of the fallen paratroopers into a C-130 transport plane. By the time I counted 12 I turned away, sick to my stomach, I couldn't watch any more.

LTC Hendrix was a tough no-nonsense Infantry Officer somewhat reminiscent of George C. Scott's portrayal of Patton. He also reminded me for all the world of Daddy Warbucks in the movie Annie. He had close-cut blond hair instead of shaved bald but the look and mannerisms seemed the same. Neither of these movies had been made at the time. He commanded a great deal of respect but he had a dark side that made him difficult to like.

We were helicoptered to a hilltop somewhere in the vicinity to set up a new firebase (Hill 530). This was probably on Nov 25 ('67). The next day I was standing in front of the radios in the TOC when there was a loud blast and I felt like someone had just put his cigarette out against my butt.

I was looking toward the infantry radio operators and one of them, a man named Johnson, rolled his eyes up into his head and slowly sunk to the floor, we thought he was dead. I was a little confused about what just happened until I glanced at the ceiling and the hole the 82-millimeter mortar round had made. The enemy's first round had hit the TOC! I thought it was a well-aimed shot, not a lucky one, and that next one would finish us off. I wanted to run but where? In Vietnam there was nowhere to run. The mortar round had also put out the main lights in the TOC. The back-up lights were working but were very dim. My boss, Captain Hill (6th/29th Artillery Liaison Officer), was off visiting over at (B Battery) FDC, he was always visiting somewhere, so Ken Marshall and I dug out a flashlight, cleared the cannons and called up the artillery batteries to shoot them. The mortar rounds were still raining down on us and we were taking a lot of casualties. Captain Hill eventually showed up and took over. We continued to fire counter mortar fire the rest of the day into the night and the NVA for their part kept shooting at us. The firing did dwindle down though Jack Enright showed up and relieved me so I could go to the medics. They put a field dressing on my wound and ordered me to Med Evac with some other wounded. The wounded included LT Sander of B Battery and one of the FDC chart operators along with some from the Infantry. As we flew off in the Huey we could see gun flashes in the dark jungles as the gooks tried to shoot us down.

I was treated and released the next day from the MASH at DakTo. Johnson came back to us after about a month, having spent recuperation time in Japan. After a few days we were helicoptered to Hill 1338 to set up a firebase. All the hills in the Central Highlands were named by numbers that corresponded to how high they were above sea level, thus 1338 was 1338 feet above Dak To which was at about sea level. It offered quite a view. I was fascinated by exploring the NVA fortifications and coming to appreciate how difficult it had been to climb the hill in the face of the NVA's resistance.

If you consider that the "front" of 1338 faced Dak To, then the "rear" was a hill even higher. One day the NVA fired rockets at Dak To from the hill behind us, the rockets flying over us on their way. Of course, we immediately fired on the rocket position but by then the gooks had probably run away. A Huey that later flew over the rocket position to photograph it crashed. Helicopter accidents were very common in the Central Highlands. One resupply helicopter coming into 1338 crashed and burned.

One amusing thing happened at this time. We had some troops stationed on a hill nearby. One day they decided to change positions and moved out. A few hours later they heard yelling and firing from their former position. The NVA were attacking, thinking the Americans were still there. The FO called for fire and we shot up the position.

Not all our encounters were with the NVA or VC. One night one of our listening posts experienced something entirely different. A tiger activated a trip flare and in its startled confusion grabbed a black GI in his mouth and started to run off with him. He and his buddies put up enough commotion that the tiger dropped him and ran off. The frightened man kept his sense of humor, claiming the tiger must have been prejudiced. The exact same thing also occurred to another member of the 4th ID, only this time the offending animal was an orangutan.

Captain Hill decided he'd let us go to Camp Enari, and thus to Pleiku, to do Christmas shopping. This would be a sort of three- day pass. I was chosen to be first. He wasn't around when it was time for me to go so Jack wrote a 'to whom it may concern' letter explaining what I was at Camp Enari for and that I was "exempt from all details". We got Captain Foy, one of the Infantry Officers who worked at the TOC to sign it and off I went. When I showed the letter to the First Sergeant at Camp Enari (First Sergeant Ernest Ferrante) he told me that no one from the battalion was allowed in town and then made me "Chief of Smoke". When it was obvious I didn't understand what he meant he explained I would be "burning shit" from the outhouse. It was a very unsavory job but only took me an hour to do in the morning and another hour in the afternoon. Of course I didn't report for more duty when I was done but made myself scarce until mealtime. The next morning I told him I was taking the convoy back to Dak To. He said he had talked to whoever was in charge concerning my "special case" but I wasn't buying any of that and ended my "pass" early. I think the moral to that story is that you do not tell the First Sergeant you are exempt from all duty, he'll prove you wrong!

From 1338 we moved to another firebase further west. This hill had been heavily bombed and hit with napalm. We actually were able to put the TOC in a large bomb crater. Jack, Ken and I chose a spot near the cook's tent for our sleeping bunker. Between our bunker and the tent we discovered someone's foot sticking out of the ground. He obviously had been hastily buried by his buddies during their retreat. We threw a couple of shovels of dirt on it.

There wasn't much in the way of entertainment in the firebases. The military operated a radio and television station. When I was in base camp (Camp Enari) I watched television. They were playing some pilot show that never got off the ground called "Chester The Jester" starring Gene Kelley as I recall. There were some radios out in the field though and when we had the chance we crowded around them to listen to the Chris Noel Show in the late afternoon and the Gary W. Gears Show in the evening. Chris Noel was a very sexy voiced girl who played records and Gary W. Gears did the same. I had a radio I dug out of the trash and got it to work but all I could get on it were the BBC, Radio Hanoi, and Radio Peking along with a few Asian stations that never spoke in English. Before we went to the Dak To area they brought us in to Camp Enari for showers and new uniforms. Jack had a friend who was involved with the entertainment at Camp Enari and found out where they were going to have a show. So that evening we were able to watch Philippinos put on a great song and dance show. As I recall "Mustang Sally" was requested again and again. That's Mustang Sally the song I'm referring to.

A few of us were supposed to be picked up to go to the Bob Hope Christmas Show but no one from Liaison really wanted to go so the only one I knew that went was a Jew named Joel Friedman, the Major's orderly. I thought it was kind of funny. On Christmas Day, though, none other than Chris Noel came out to the Firebase. She talked to us in that sexy voice and sang us a number of songs. She was wearing a mini-skirt and was quite a knock-out. Ken and I were on duty so we took turns, running out of the TOC for one song then back to relieve the other guy for one song and so on.

There was supposed to be a cease fire for Christmas but the gooks had to start shooting mortars at one of our companies so we spent the late afternoon and evening shooting back.

GRIZZLY'S SKI LODGE

We next were sent to a firebase in the middle of nowhere on one of the highest hills they could find. This firebase was way up in the clouds (Note: 1 mile high). I mean that literally, the clouds were often below us. It got freezing cold up there and we were all given field jackets, something almost unknown in Vietnam. Besides bringing in Bravo Battery's 105's, we also had a battery of 155's from the 1st/92nd Battalion. They were equipped with M-114 155 mm towed howitzers. They had a much longer range and could hit just about anything from this elevation. This was one of the few firebases we had that were guarded by South Vietnamese soldiers instead of our own troops who were all in the boondocks looking for the enemy. The name? LTC Hendrix's radio call sign was "Grizzly", hence "Grizzly's Ski Lodge". No enemy was found and there was nothing for the artillery to shoot, so after a week or so we moved. After about three quarters of everything was helicoptered out and we had destroyed our bunkers, the helicopters were called away on urgent business. Jack and I, along with a platoon or so of Infantry, were stuck on the hill for the night without anything to eat. We made a makeshift bunker out of ammunition crates and searched through the garbage dump for unopened cans of C-Rations. I found a can of ham & lima beans. We all hated ham & lima beans but in this case I thought it was delicious.

FSB 25

Down the road from Dak To toward the Cambodian border the Special Forces were busy setting up a stronghold called Ben Het. We were sent to a nearby hill to support their defense (FSB 25). Units of the 101st Airborne were holding the hill when we got there. We were taking over. I was on radio watch that first night. The paratroopers kept throwing hand grenades out into the jungle. They called them H&I but it seemed to me they were scared of their own shadow and it kept us feeling jumpy all evening. In the morning a dead paratrooper was found entangled in the concertina wire outside the firebase. He had been part of a listening post and when they came back into the base he got tangled in the wire and someone threw a hand grenade at him. Officially it was recorded as an enemy mortar round that killed him. Ken Marshall had been part of the Advanced Party that went to the firebase the day before the rest of us. They had been attacked that first night with the paratroopers and he was still pretty shook up when we saw him. No one from Liaison again went with the advanced party.

The Fourth Division participated in what was called an Infusion Program. This was a switching of personnel from one division to another. This would prevent the situation of the previous September when the Fourth Division found itself with all new guys. Jack and I were both on the list to go. Captain Hill called somebody to see if he could get me off the list but things didn't look too promising. It was still a way in the future though. Major Stiner was reassigned and replaced by Major Bradshaw as 3/12th Inf Battalion Operations Officer. FSB 25 came under mortar attack many times while we were there and the TOC had taken many more direct hits but it had been made "proof" against the communist 82mm mortar and they didn't get through to us.

Unfortunately 82's were not the only thing the NVA had available to them. One evening our Forward Observer, 49 (LT Joe Knowlton) called us on the radio to say the enemy was shooting at someone from close to where he was. At the same time we started receiving mortar fire so it wasn't hard to guess who the target was. These weren't the standard 82's though, but 120-millimeter mortars! They were much more powerful and quite capable of penetrating any of our bunkers. We all put on our helmets and flack vests but didn't feel safe. One of the infantry radio operators took the metal top off a file box and held it over his head, nobody laughed. 49 kept calling to let us know "Three on the way!", "Four on the way!" and so on. We nervously counted the explosions and breathed a sigh of relief when the last one hit. Eventually the mortar fire stopped, or our counter mortar fire stopped it.

The enemy also had a battery of 100-millimeter Russian guns they started shooting at us with from inside Cambodia. These also could take out any bunker they hit. We couldn't tell exactly where they were firing from. We needed observation aircraft in the area when they fired to spot them, and since they were firing from Cambodia we didn't have anything in the vicinity that could reach them. The Russian 100's had a much longer range than our 105's. Luckily for us they only fired four or five rounds at a time so we wouldn't spot them. But they were a nagging problem for us for a long time. Like an itch you can't scratch.

One day Jack and I went to Dak To for some reason I've long forgot. After we got out of the Huey and were walking away from the helicopter pad, we heard that old familiar sound of incoming rounds. Dak To was being rocketed! We immediately ran to the nearest bunker, it was small like a fighting bunker and very solid and strongly built. Another trooper also jumped inside with us. After a minute or so he declared the bunker was "too small to withstand a hit" and ran out to a bigger bunker somewhere else. Jack and I both laughed out loud. We were in the safest place you could be. The large community bunkers they were starting to build weren't necessarily stronger and they were easier to hit.

Whenever we went to Dak To we usually stopped off to visit with Brigade Liaison. They were situated in a big command bunker we had first used when we came to Dak To. We had a radio tuned to the Brigade frequency at all times and talked to these people every day. I only remember three of them now. Captain Sammons was the Liaison Officer, and two radio operators, a black Sergeant, and John Parent. Their radio call signs were 22, 22 A, and 22 C respectively. 22 A had a bad speech impediment and was difficult to understand. When Brigade had business for us it was often for the various Battalion Liaisons. They frequently had 22 A repeat his messages many times. But I got so I understood him pretty good and when he asked "How do you copy?" after a transmission I would jump in first and say "I have good copy but will read back for possible correction". I'd then read it back and everyone would get it second time around. Captain Sammons seemed to especially like me and asked me to come work for Brigade but I declined politely. I wanted to be where the action was.

Three of the infantry radio operators that worked in the TOC were Bill Hurst, Merlin Meadows, and Jim Birge. Bill was 23 (old to me) and on his third marriage. Jack Enright's sister had a connection with a Catholic Girls school near Chicago called "Immaculata" as I remember. Anyway, they wanted to correspond with servicemen in Vietnam, as part of their patriotic duty I suppose. Jack's sister sent all these letters to Jack to distribute. Bill started to write to one of the girls and pretty soon her sister and I think even her mother started to write to him and send food packages. To make a short story shorter, I don't know what sort of BS Bill was writing but the girl and apparently her family were soon thinking of him as a potential husband for her. He was just a smooth-talking con artist.

Merlin Meadows was from Garden City, Michigan. For many weeks we had the night radio watch together. He wrote short stories and poems that the Garden City newspaper sometimes printed. One night we decided to sing a number of songs in a low voice so we wouldn't wake anyone up. The next morning the Sergeant Major complimented our singing but told us "No more" because we kept him up most of the night.

Jim Birge was a kind, gentle soul that had no business ever going to war. He had a terrible time keeping awake at night and I'd sometimes keep an ear for his radio as well as my own.

Jim and Bill raised the ire of LTC Hendrix for failing to report some "important" message they had received over the radio. He sent them out into the field with the companies. Bill was involved in a fire fight with the enemy and was seriously injured by a hand grenade blast. He recovered and came back to work with us in the TOC.

One day one of our Forward Observers, 48, called for a recon by fire. This was a very common practice of shooting artillery ahead of yourself to break up any possible ambush before you walked into it. I marked 48's position on the map, the spot he wanted to shoot and put a ruler from the gun-sight to the target. 48 was in the line of fire if we fired from B Battery's position, unsafe. I advised 48 and told him we could fire it from another battery's position in safety and with room to adjust the fire if the enemy was there. I even called up the other battery with the fire mission. 48 was new, our 48's were always new, and he wanted to fire from B Battery because he knew and trusted them. Our Liaison Officer wasn't there and Major Bradshaw, the new 3rd/12th Battalion Operations Officer got into the act. He yelled at me that I had no authority to refuse fire to my forward observers. I was scared to death! I had never stood up to an Officer before. This was the Army and you did as you were told. I had visions of being put in front of a firing squad, but I also had visions of American troops being blown apart because of me. In as polite and respectful way as I could I told the Major that was the very reason for having a Liaison, the mission was unsafe, but I would let it be fired if he signed the log stating it was fired on his authority. I was shaking like a leaf. Major Bradshaw gave in. Much relieved, I told 48 he'd have to use another battery. He decided not to shoot the mission now.

Captain Hill was transferred to Brigade to take over from Captain Sammons. Captain Lambert replaced Captain Hill. Captain Lambert was a very likeable, easy going person filled with self-confidence, similar to Major Stiner. He also knew his job from the start.

We also received two new radio operators, ostensibly to take over when Jack and I were infused to another division. One was Norman Tolly who had been transferred from the Americal Division and was experienced in Liaison. The other was Glen Trimble, green as grass.

One morning Glen was shaving and brushing his teeth in front of our sleeping bunker. I was standing next to him talking having sort of a bonding session as it were. I heard that old familiar sound of an incoming 100-millimeter round. I ducked, Glen didn't and caught a piece of shrapnel in the jaw. Both of us threw our ourselves on our bellies and did a fast crawl into the bunker. The first shell had hit very close and they kept coming in. Glen wasn't really hurt too bad and our main concern were those rounds that kept screaming in at high speed looking to kill someone like us! The shelling soon ended, not soon enough, and we emerged from the bunker to find Glen's tooth brush had been snapped in two by a shell fragment, probably the one that hit Glen in the jaw. Casualties of war.

They brought a radar unit to the firebase to try to locate where the 100-millimeter battery was firing from but the Russian guns fired at too low a trajectory for them to get anything but a rough estimate. We were still getting hit fairly regularly with mortars though and the radar unit was very good at pinpointing their position.

One evening we were being hit with mortars and they somehow set one of our ammunition bunkers on fire. Someone came into the TOC and asked for volunteers to help fight the fire before the bunker exploded. I hadn't done anything but my job since I'd been in Vietnam, nothing over and above that is, so I volunteered. It was pretty eerie out there carrying water to the burning bunker and listening for the hiss of an incoming mortar round, all in the dark, but we got the fire put out. I even ran into Calvin Thompson that night. He had been in the 6/92 Artillery back at Fort Hood with me. I knew that he was with one of the gun crews but this was one of the first opportunities we had for a talk and we stayed out talking after the fire was out. The mortar fire had dwindled down to a round exploding every once in a while by then.

One of our Infantry companies went out on patrol. We usually had one company at the fire base at all times and the other three out on patrol. This time they made heavy enemy contact right away. The NVA may have been massing for an attack on the firebase. Anyway, the enemy got too close to use artillery effectively. It became a running gun fight. The Americans running, the NVA gunning. The Americans fell back to the firebase with the gooks close behind. The gooks were shooting mortar rounds into the clearing between the firebase and the jungle. All the TOC radio operators who weren't needed in the TOC were sent to help bring in the dead and wounded. The whole thing was one hell of a mess. The gooks, in their eagerness were actually breaking the cover of the jungle to come out in the open to shoot Americans. They were in turn shot by the machine guns in our perimeter bunkers. Jim Birge was with that Infantry Company, reaching the clearing but a mortar round killed him.

This was a very hot area to be in now. The NVA was desperate for a victory and they seemed to have chosen us. The Air Force flew B-52 strikes in the area to try to break up the enemy forces. This was a type of carpet bombing of an area, we called them "Arc Lights". They were quite a sight to see in the daytime, throwing up huge "curtains" of dirt and dust clear into the sky and rumbling the ground for miles around. They also flew WWII era C-47 cargo aircraft converted to gunships that fired dozens of Gatling machine guns at such a volume and rate it looked like a stream of light coming out of the black night sky to the ground. But it was actually thousands of tracer rounds. From where we were it sounded like a loud, elongated fart!

Several times the NVA massed together outside the firebase in preparation of overrunning us. When they did we fired artillery from other firebases and also fired "beehive" at them. Beehive was a round fired by our 105's that exploded soon after leaving the 105's tube and throwing thousands of "darts" in a shotgun pattern. These "darts" were actually short finishing nails with fins molded or stamped on the "head end" to make them fly straight. They could easily go straight through a person and were very effective in breaking up massed attacks.

A number of times we called for close air support. These were sometimes delivered by jet fighter-bombers, "fast movers" as they were called, and sometimes by propeller driven Sky Raider Dive Bombers. Anyway, one day a number of gooks set up a machine gun in a bomb crater some 25 meters outside the firebase. We called an airstrike on them. I laid on the ground watching as a Sky Raider dived on the bomb crater and saw the bullet tracers shooting up at the plane as the Sky Raider shot back. The Sky Raider won.

One day I took an urgent call from Forward Observer 47 (LT Bill Davis) asking for artillery support because his lead elements were in contact with the "enemy". At the same time 49 (LT Knowlton) was calling with the same request. One glance at the map and I realized they were in contact with each other, and I put a stop to it, luckily before anyone got hurt. I had been monitoring their movements all along but had assumed the Infantry would have kept each company apprised of the other's location. From that point on I assumed that task myself and on at least two other occasions managed to prevent a similar occurrence.

We weren't the only ones who had that problem. One night I heard a lot of shooting from an area I was sure we had no troops. I ran to the TOC to check for sure. I was right, the shooting must have been coming from two enemy elements that "bumped in the night".
Enemy fire wasn't our only concern. One night as I lay asleep I came out of my slumber by the unmistakable sound of an incoming artillery round. It hit the firebase with an especially heavy explosion. I was wide awake and alarmed now. This was followed by two more of those heavy shells. It turned out it was a 155-battery shooting a mission without our clearance across our area of operations. This was supposed to be cleared by us but mistakes happen. Jack recognized it for what it was, called Brigade Liaison which in turn halted all firing until it could be sorted out. No one was hurt. Lieutenant Ray Carpenter, Executive Officer of B Battery, told me he was outside when the first round hit. He ran to one of the artillery bunkers, the blast blowing the wall in on top of him.

It seemed that we finally broke the siege of Ben Het and it was time to move. As the helicopter came in to take us and our equipment out they threw up a lot of dust and we instinctively ducked down. One of B Battery's cooks was a young black man named Green that everybody loved. He ducked next to the pile of excess powder stored from the artillery rounds prior to burning them off. A spark thrown up, or out, by the helicopter ignited the powder in a big flash. Green was terribly burned and had to be Med Evaced. Good riddance to firebase 25.

1338 REVISITED

Ken went on R&R (Rest & Recuperation) to Hong Kong and we went to Hill 1338 overlooking Dak To. It was being called R&R Hill now because nothing much happened there anymore.
We usually maintained a small firebase some distance away from us. In this firebase we had one or two mortar tubes. The mortar's call sign was 72, so the remote firebase mortar's call sign was

72 R, "R" for remote. One night we got a call from this firebase to inform us the rain had washed out one of the walls from one of their bunkers and the roof caved in. The three soldiers inside were crushed to death! Senseless accidents were a big problem in Vietnam.

Dak To had always been a target for frequent rocket attacks. These were 122-millimeter rockets similar to, or exactly the same as those the Russians used against the Germans in World War II to great effect. They had about the power of our 105's but they were not very accurate. 22 C, John Parent, told me that just about every time Dak To was rocketed, someone got killed, so they couldn't be ignored.
To the left of 1338 as you looked at Dak To was a large ridgeline that also overlooked Dak To, though from a lower height. Our intelligence got word that the NVA were about to shoot rockets at Dak To from this ridgeline. We helicoptered troops down to the ridgeline to prevent them. They immediately ran into a strong enemy force and a fierce battle ensued. From our vantage point on top of 1338 we could hear the gunfire and see the helicopter gunships making their firing runs in support of the troops. Even while the battle was going on, the enemy started launching their rockets. Helicopters made firing runs at the sites but still the rockets launched. I'm sure our efforts prevented all the launchings planned by the NVA but the fact that any launchings occurred at all is a powerful testimonial to the tenacity of the enemy.

Dak To continued to be attacked. One night a huge explosion occurred that actually seemed to shake 1338. I ran out of the TOC to see a giant mushroom cloud rising out of Dak To to thousands of feet above me! It looked like Dak To had been hit by an atom bomb! Actually an 8-inch howitzer battery's ammunition bunker had been hit.

Ken came back from R&R and some weeks later it was my turn. When I arrived back at Camp Enari Jack Enright was there in the transit barracks. While I was in Hong Kong the 4th ID had called for those on the infusion list for transfer to other divisions. Jack told me I was on the list and should report to the First Sergeant in the morning. I knew the Army, though, if I hadn't been ordered by someone in authority to do something I could hardly be held responsible for not following orders. The next morning, before the morning formation, I joined the convoy out to Dak To. Before the First Sergeant could read the log and see I was back, I was already gone. I don't know if Captain Lambert actually believed I didn't know I was supposed to be infused; I told him that I hadn't talked to anyone but had automatically joined the morning convoy. Anyway, he was glad to have me and nothing more was said.

FSB 29

We put troops on Hill 824, south of Ben Het, with the idea of setting up a firebase. The NVA were desperate to get us off the hill. They attacked in force getting through the perimeter defense and actually taking over two or three of our perimeter bunkers! We had to use flame-throwers to get them out. I was safe on 1338 that night, listening to the developments over the Infantry radios in the TOC. We were to move to that hill after daylight to set up a firebase which would be called Fire Support Base 29. The move to FSB 29 was pretty uneventful but now began the task of clearing the area and sandbagging the TOC in. This consisted of building a bunker around the radio shack that consisted of a conference room just outside the door of the shack with a room on either side, one for officers to sleep in and one for the sergeants. This took pretty much all day.

Captain Lambert put me on night shift after I got back from R&R. The night man worked alone and pretty much had to be able to handle anything that came up by himself. Tolley could have handled it but I think Captain Lambert wanted me "kept out of sight" as it were, until the infusion thing blew over. Anyway, I was thinking of a place to sleep before my shift began and was looking over an abandoned foxhole. Ken walked up and told me that was where Meadows, our singer, poet, short story writer, was killed the night before. He had come here as Advanced Party. We had so much to get done before night-fall, when we expected to be attacked again. We didn't have time to grieve for him, and pretty much put him in the back of our minds.
Circumstances being what they were, it was decided to split the night shift so everyone could get some sleep. All the radio operators would sleep in the TOC that night. By the time I went to find a place to lay down, the one dry spot I could find was only big enough to keep my upper trunk dry. From the waist down I slept in the mud.

Firebase 29 was on a hill with two ridgelines, one projecting at a right angle near the east end. We occupied the ridge that ran east to west. The TOC was located by the far east end, just beyond the north-south projection. The night before the gooks had used the north-south projection to overrun the three perimeter bunkers, allowing them to avoid fighting uphill.

Ken woke me for my shift in the wee hours of the morning. What we were doing was shooting up the north-south ridge at staggered times in order to break up any massing for attack. This was being fired "Danger Close" from a 155 battery at another firebase. "Danger Close" meant that artillery was shooting so close to friendly troops that each gun had to be personally inspected by an officer to make sure it was set right before shooting. Before shooting the firing battery would have to clear through me. So this is the way the night went: "40 C, this is Upland 43, ready to fire, over". I would answer "Upland 43, this is 40 C, wait over". I would then have the Infantry operators alert the perimeter to keep their heads down and call back "Upland 43, this is 40 C, shoot it, over". "This is Upland 43, shot, over". "This is 40 C, roger, shot out". This went on all night long. In the morning five or six dead gooks were found on the ridge with fresh blood trails that showed others had been dragged off. I watched that morning as the Infantry buried them, carrying them with sand bags over their hands so they wouldn't have to touch them.

After breakfast Tolley and I went to look for a place to build a sleeping bunker. We found a spot in the middle of the ridge. Some Engineers were there with a back-hoe digger so we had them dig us a nice deep hole and set about covering it with tree trunks and sandbags. Suddenly we started taking fire from those damn 100's. We ran to a nearby bunker taking advantage of a trench that had been dug across the ridge and waited it out. We finished building our bunker. The Infantry cooks built a bunker right next to ours, using the dirt the back-hoe had dug out of our bunker. They built a nice big one, above ground, and added insult to injury by putting up a big American Flag. Over the next several weeks we received many close rounds of enemy fire and I couldn't help but blame that damn flag that made such a good aiming point. One does not want to stand out in a war zone.

Out to the west of us, toward the Cambodian border, along the road that ran past Ben Het was an old French Foreign Legion Fort. We sometimes used it to register our guns on. It sort of brought out the romantic in me. I always had a fantasy of joining the Foreign Legion. The irony being that if I had joined fifteen years ago I could be just about where I was now, serving when the French were there in the 1950's.

To the west of us, slightly to the north, was a hill that was roughly as high as ours. Several times the NVA set up a recoilless rifle on that hill and shot at us. A recoilless rifle was a direct fire weapon that you aimed right down the barrel at what you wanted to hit, just like a rifle. The ones the NVA used were 75-millimeter, the same size round as fired by a Sherman tank. Recoilless rifles were especially terrifying since they gave little warning. You heard what sounded like someone popping a cork off a champagne bottle in the next room, followed by a second of silence, then, "Sa-Bang!" You have no time to react once you hear that round screaming in. By the time you hear it, it's either going to kill you, or it isn't, and it always sounds like it is.

One day I was standing outside our bunker talking to Tolley when one of those rounds screamed in and hit between our bunker and the cook's bunker. We practically flew into our bunker and then began one of the most terrifying three or four minutes of my life. The gooks fired several more rounds at us, each hitting spitting distance from us, one over, one short, one over, one short. It was obvious that we were the ones they wanted to kill and no one else! We had been standing out in the open in plain view of anyone watching from that hill with the recoilless rifle. Right near that big flag! After what seemed like an eternity, they stopped firing, no doubt they picked up their gun and ran with it before we could shoot back. We breathed a big sigh of relief. Two days after I left Firebase 29, our bunker took a direct hit and the roof caved in. Luckily, no one was in it.
I always considered Firebase 29 to be the most dangerous of all our firebases. We were always surrounded by large numbers of enemy and the menace seemed to be a physical thing that filled the very air we breathed. People seemed to be getting killed all the time and it was a matter of time before it was my turn.

Most of the events there are long lost to memory. I do remember one three-day period when we were under continuous shelling day and night. Sergeant Williams (Mad Willie) looked at me during some particularly heavy shelling and said almost to himself "Jesus, two weeks". What he meant was that he had two weeks left before returning home and he might die here today.
One of the radios stopped working and it was obvious that the cable to the antenna had been cut. It was pitch black so I was feeling along the cable, "I think I found it" I declared. In a much louder and assertive voice the Sergeant Major standing behind me said "There it is right there". I looked to where he was shining his flashlight and saw a complete break in the cable, two feet from where I was holding my thumb over the gouge. This brought howls of laughter.

But still the shelling continued. This was not continuous but intermittent. It wore heavily on our nerves and I developed a splitting headache. I happened to be passing by Mad Willie and must have given him a look that conveyed all I felt because he answered my unspoken statement with "I know, I have a headache too". Whether it was our counter-fire, aerial attacks, or they just gave up or ran out of ammunition, the bombardment stopped.

We finally got the 100-millimeter battery. It started firing on someone and the radar unit got a fix on it. An aerial observer was quickly overhead even though it was in Cambodia and directed the heavy American artillery we had specifically brought in to deal with them. The NVA tried to save their guns by moving them out of harm's way, but the American observer was too good and all the guns were destroyed along with many of the cannoneers.

Captain Lambert left us and was replaced by Lieutenant Parsons who had been our Forward Observer with D Company. He was the one we knew as 50. This was not a good time to be new at your post. The NVA was reinforcing the area with their regiments with orders to wipe us out. That physical air of menace was hanging over us again and we expected all hell to break loose at any time.

This was the end of our tours, and Ken was ordered to Camp Enari for processing home. Two days later it was my turn. I have to admit I was not that excited about leaving. I certainly didn't want to die but I was on the adventure of my life and I knew it. "The World" as we called the United States seemed a distant memory: Vietnam was a reality I knew, but I'd already given the cylinder a spin, put the gun to my head and pulled the trigger on an empty chamber too many times. It was bound to be loaded one day so the smart thing to do was to quit while I was ahead. I made my rounds to say goodbye, got on a chopper to Dak To and then the afternoon convoy to Camp Enari.

Ken was still in the transit barracks when I got there. We had been together for a whole year, working side by side for some sixteen hours a day, sleeping in the same bunker, two or three air mattresses side by side just high enough to sit up in. On the Ski Lodge we'd wrap ourselves in our own blanket and pull up Jack's blanket over the two of us. At the beginning we made a pact to watch each other's back and that's what we did. When I went down to Dak To for whatever reason he'd figure out what time I'd be waiting for a helicopter lift back to the firebase and call the loadmaster and tell him I was a high priority so I'd always go back on the next chopper no matter who was waiting or for how long.

I also saw "22 A" there. He was on his way home too. He was a Sergeant and had a private room and invited me over for a drink of scotch. Sometime earlier I had run into 22 C, John Parent, in Camp Enari and the two of us had gone out to some bar as I recall. These two were with Brigade Liaison, people we talked to on the radio every day.

The biggest thing for me was that I'd get to see my childhood best friend, Leonard Stone. He was a clerk, or secretary, for the 4th ID and processed the new arrivals. I had hoped to see him on my last leave but he had been held back in Basic Training. We had two or three evenings together and then I was off to Cam Ranh Bay and my flight Home.

I was only in Cam Ranh Bay one or two days but I was put on, of all things, KP duty. I hadn't been on KP duty since Oakland when my last duty in the States was to mop floors. And my last duty in Vietnam? Mopping floors!

We were all dressed up in our khaki uniforms waiting for the bus to the airport. We were all in good spirits. We were going home! Suddenly one Sergeant let out a moan, fell on the ground and went into convulsions. A couple of guys held him down so he wouldn't hurt himself and someone ran for the medics. He was hauled off on a stretcher, another man was brought over quickly to take his place. As the wicked witch said to Dorothy, "Leaving so soon? I wouldn't hear of it!" Vietnam!

The flight was not uneventful. Something went wrong with the plane and we landed in Tokyo. The airline took us to a large hotel where we ate lunch and loitered about for several hours watching Japanese television. The Japanese Army was fighting pre-historic dinosaurs, it was all in Japanese of course, but still interesting. None of us had watched television for a year. Then it was off to Fort Lewis, Washington, a steak dinner, another day of processing, and home. This was August 1968. I was twenty years old.

In looking back I have a lot to be grateful to a lot of people for my survival. First would be to my Uncle Murray for advising me to do my very best on the examinations. I did and it eventually took me to Artillery Liaison. To all those I came in contact with who kept me laughing through some very dreadful times and who did their jobs and made everything work. From the Officers to cooks to the supply sergeants, they were all important. The "coulda beens" are endless so I have to be grateful to the Grace of God which saw me through alive. In all the thousands of artillery fires we cleared no American troops were killed by the first rounds, the ones that were our responsibility. I heard that another Liaison section killed twenty-five in six months. I can only thank God for that too. We did make mistakes, but no one got hurt as a result.

THE UNIQUE VIETNAM EXPERIENCE

As I reminisce about Vietnam, the memories that are unique ONLY to the Vietnam experience come to mind. "The World". That meant "Home", the place we had come from, and hopefully were going back to. And to get there, we would be flying on a "Freedom Bird". One of the ironies of the Vietnam War was that you could be minutes from a mortar attack or fire fight on the ground, and look up and see a US commercial airliner flying overhead, presumably bringing fresh replacements "in country", and then going back to "The World" with lucky Vietnam Vets. About a month before DEROS (date you were scheduled to depart Vietnam at the end of your tour), many Soldiers started sporting a "short timer's calendar", usually mounted on a clipboard with a drawing of a pin-up girl with a "by the numbers" puzzle "fill-in". Of course, the ultimate goal was to fill in that "piece" that was number 1. To get to that point, one went from being a "2-digit midget", to a "1-digit midget" with 9 days or less remaining "in country". The final 3 puzzle pieces, days 3, 2, and 1, were the choice areas of interest for any short timer getting ready to return home. I leave it to your imagination to remember what those were! Before boarding the Freedom Bird, there was the "Duffel Bag Drag" to the air base, usually flying from Cam Ranh Bay.

And so it was for me in early September, 1968, in Cam Rahn Bay, wearing a khaki uniform again, ready to get on a flight home. Every soldier there was in a happy mood, anticipating the long awaited moment for their turn to get on the "Freedom Bird"! Soldiers came from many different units. As a way of showing off our individual unit pride, we were given plastic Division patches to hang on the front pocket of our khakis. I proudly wore my 4th ID patch!

As a final thought, I want to share LTC Albert Akers' advice to the Leaders in the Battalion: "Remind your Soldiers that although most do not realize it now, they will look back on their year of service in Vietnam as the most memorable event in their lives." I couldn't agree more!

BIBLIOGRAPHY

After Action Review (AAR), 3rd Bn/12th Inf with following dates: 20 Jan 1968, 29 Jan 1968.

Buchanan, Donald W., Staff Sergeant (Later First Sergeant), Chief of Section (Base Piece), B Battery, 6th Battalion, 29th Artillery, 1st Brigade, 4th Infantry Division (1967-1968). Provided personal experiences through many conversations at several reunions included in this book. (2010-2019).

Campbell, Louie L. (Pee Wee), Private 1st Class, Cannoneer, B Battery, 6th Battalion, 29th Artillery, 1st Brigade, 4th Infantry Division (1967-1968). Provided photos and personal experiences through many conversations included in this book (2010-2018).

Carpenter, Ravin R. (Ray), 1st Lieutenant (Later Captain), Executive Officer, B Battery, 6th Battalion, 29th Artillery, 1st Brigade, 4th Infantry Division (1967-1968). Provided personal experiences through many conversations included in this book (2010-2018).

Dak To and the Border Battles of Vietnam, !967-1968 by Michael A. Eggleston, ISBN 978-1-4766-6417-0, McFarland & Company, Inc., Publishers, Jefferson, NC, 2017.

Daily Log (DA Form 1594), 6th Bn/29th Artillery with following dates: 20 Jan 1968, 27 Jan 1968, 2 Feb 1968.

Davis, William P. 1st Lieutenant (Later Captain), Forward Observer, B Battery, 6th Battalion, 29th Artillery, 1st Brigade, 4th Infantry Division supporting 3rd Bn/12th Inf (1967-1968). Provided photos and personal experiences through many conversations included in this book (2008-2018).

Farris, Samuel D., Jr., Specialist 4th Class, Cannoneer, B Battery, 6th Battalion, 29th Artillery, 1st Brigade, 4th Infantry Division (1967-1968). Provided photos and personal experiences through many conversations included in this book (2010-2014). Earned Purple Heart for wounds sustained during Battle of Dak To on Hill 530 late November 1967. Purple Heart Medal not received until 47 years later thanks to combined efforts by Specialist Farris and his Battery Commander on Hill 530, CPT Michael George.

George, Michael S., Lieutenant Colonel (Ret.) – Captain, B Battery Commander, 6th Battalion, 29th Artillery, 1st Brigade, 4th Infantry Division, 1967. Captain George and the Military Order of the World Wars (MOWW) gave permission for inclusion of the following article authored by Captain George printed in its entirety in this book: ***Thirty Days At Dak To—A Field Artillery Story*** by LTC Michael S. George, Officer Review, Volume 48, Number 6, The Military Order of the World Wars, January/February 2009.

Gillen, Nickolas J. Sergeant, Cannoneer/Gunner, B Battery, 6th Battalion, 29th Artillery, 1st Brigade, 4th Infantry Division (1967-1968). Provided photos and personal experiences through many conversations included in this book (2010-2016).

Haas, Ronald K., Specialist 5th Class, Fire Direction Center (FDC) Chief, B Battery, 6th Battalion, 29th Artillery, 1st Brigade, 4th Infantry Division (1967-1968). Provided personal experiences through many conversations at reunions and shared correspondence and emails included in this book. (2008-2018).

Howitzer Happenings, 6th Battalion, 29th Artillery Newsletter, Number 7, 15 June 1968, article titled ***Redlegs Become Infantry***, author name not included.

Kenney, Donald E., Colonel, US Army (Ret.) – Lieutenant Colonel, Commanding Officer, 6th Battalion, 29th Artillery, 1st Brigade, 4th Infantry Division (1967-1968). *Operational Report for Quarterly Reporting Period Ending 31 January 1968.* Personal correspondence (2014-2016).

Manny, Larry G., Specialist 4th Class, Cannoneer, B Battery, 6th Battalion, 29th Artillery, 1st Brigade, 4th Infantry Division (1967-1968). Provided photos and personal experiences through many conversations included in this book (2012-2014).

McEwen, Howard D. (Tappy), 1st Lieutenant, Forward Observer, later Fire Direction Officer, B Battery, 6th Battalion, 29th Artillery, 1st Brigade, 4th Infantry Division (1967-1968). Provided information from his Vietnam scrapbook and personal experiences at reunions included in this book (2008-2010).

McNulty, Patrick J. Jr, 1st Lieutenant, C Battery Forward Observer with 3rd Bn/ 8th Inf, later B Battery Forward Observer with 3rd Bn/12th Inf, 6th Battalion, 29th Artillery, 1st Brigade, 4th Infantry Division (1967-1968). Provided input of personal experience and photo as a Forward Observer for this book. Many email exchanges (2014-2019).

Nicolson, John C., Specialist 4th Class, Battalion Liaison Section with 3rd Bn/12th Inf, Headquarters & Service Battery, 6th Battalion, 29th Artillery, 1st Brigade, 4th Infantry Division (1967-1968). Provided input of personal experience and photos of his year in the Liaison Section for inclusion in this book (2014-2015).

Nixon, Austin David (Dave), Colonel (Ret.) – 1st Lieutenant, Forward Observer, B Battery, 6th Battalion, 29th Artillery, 1st Brigade, 4th Infantry Division (1967-1968). Provided input of personal experience and photo as a Forward Observer for this book. Many email exchanges (2012-2019).

Shadow Warriors, Inside the Special Forces by Tom Clancy with General Carl Stiner (Ret.) and Tony Koltz, ISBN 978-0-425-18831-6, Berkley Books, New York, 2002.

Snyder, Charles A., Specialist 5th Class, Communications Section, B and Headquarters & Service Battery, 6th Battalion, 29th Artillery, 1st Brigade, 4th Infantry Division (1967-1968). Provided photos of scenes around Dak To that are included in this book (c. 2014).

Stark, Stephen T., PFC, later Specialist 4th Class, Reconnaissance and Survey Section, Headquarters & Service Battery, 6th Battalion, 29th Artillery, 1st Brigade, 4th Infantry Division

(1967-1968). Co-author of this book providing input of personal experience and photos of his year in Vietnam. Extensive collaboration (2006-2019).

The Tet Offensive, A Concise History by James H, Willbanks, ISBN 978-0-231-12841-4, Columbia University Press, New York, 2007.

They Marched Into Sunlight, War and Peace, Vietnam and America, October 1967 by David Maranis, ISBN 0-7432-6104-6, Simon & Schuster Paperbacks, New York, 2003.

Williams, James A., Specialist 4th Class, Cannoneer, later Chinook load prep team at Dak To Airstrip, B Battery, 6th Battalion, 29th Artillery, 1st Brigade, 4th Infantry Division (1967-1968). Provided input of personal experience through numerous conversations for this book (2010-2018).

Wright, Chris D. (Pineapple), Private 1st Class, Mess "Sergeant", later member of Forward Observer Team, B Battery, 6th Battalion, 29th Artillery, 1st Brigade, 4th Infantry Division (1967-1968). Provided input of personal experience for this book at Las Vegas Reunion (2012).

Wright, Richard H., Lieutenant Colonel (Ret.) – Captain, Battalion Supply Officer (S-4), Headquarters & Service Battery, later B Battery Commander, 6th Battalion, 29th Artillery, 1st Brigade, 4th Infantry Division, 1967-1968. Provided recommended reference publications for this book and details on PFC Jack Marinacci's award of the Distinguished Service Cross included in this book (2016-2019).

Tempel, John F., Specialist 5th Class, Battalion Mail Clerk and Classified Message Security Officer, Headquarters & Service Battery, 6th Battalion, 29th Artillery, 1st Brigade, 4th Infantry Division (1967-1968). Provided input of personal experience for this book at Nashville Reunion (2012).

INFORMAL DEFINITIONS, ACRONYMS AND TECHNICAL DETAILS

The following is provided to help in the understanding of life in the Vietnam Central Highlands in 1967 and 1968, and to offer a few technical details and bits of information that may be of interest to some readers. All provided courtesy Stephen Stark.

A

after-action report – A military report prepared in the field to document and summarize a specific operation, occurrence, or timeframe that is usually prepared by a field grade officer but sometimes by others.

Air America – The CIA-controlled airline that supported American interests in South Vietnam and elsewhere primarily in Asia during the Vietnam War. Air America served many purposes including delivering all types of cargo and personnel while flying both overt and covert missions.

Air Medal – The Air Medal is earned by an individual for single acts of heroism or meritorious achievement while participating in aerial flight. When earned for valor, a small "V" is attached to the ribbon.

AIT – Abbreviation for Advanced Individual Training, the school most soldiers attend usually following basic training to study their assigned MOS.

AO – Abbreviation for Area of Operations. A geographic area of responsibility for a military unit.

Army Commendation Medal (ARCOM) – The Army Commendation Medal is the fifth most significant medal earned by an individual for service or valor. When earned for valor, a small "V" is attached to the ribbon.

Artillery (Field Artillery) – The combat branch of the United States Army originally formed on 17 November 1775, now headquartered at Fort Sill, Oklahoma that employs short-range and long-range howitzers and guns (sometimes rockets) for the primary purpose of supporting infantry and armor in the field and destroying targets.

Artillery: Howitzer, 105mm – Field-portable artillery that is usually towed and capable of firing many types of roughly 4-inch diameter projectiles including high explosive (HE), white phosphorous, smoke, illumination, flechette, and COFRAM ("Firecracker"), from semi-fixed rounds. The range of a 105mm howitzer in Vietnam was approximately 11,000 meters (6.8 miles).

Artillery: Howitzer, 155mm – Field-portable artillery that is either towed or self-propelled and capable of firing many types of roughly 6-inch diameter projectiles including high explosive (HE), white phosphorous, smoke, illumination, flechette, and COFRAM ("Firecracker"), from separate-loading rounds to a maximum range of approximately 14,600 meters (9 miles).

Artillery: Gun, 175mm – Field-portable artillery that is self-propelled and capable of firing many types of roughly 7-inch diameter projectiles to a maximum range of approximately 33,000 meters (20.5 miles).

Artillery: Howitzer, 8-inch – Field-portable artillery that is self-propelled and capable of firing many types of roughly 8-inch diameter (203mm) diameter projectiles to a maximum range of approximately 16,800 meters (10.4 miles). During the time we were in Vietnam, the eight-inch gun was known to be the most accurate.

Artillery Liaison Specialists (relevant to the First Brigade, 4th Infantry Division in the Central Highlands in 1967and 1968) – A team comprised of three to five enlisted men, one NCO, and one officer that supported the planning and execution of artillery and infantry operations and were responsible for reliably tracking the location and movement of known or suspected enemy forces and all friendly elements on the ground including regular infantry, LRRP, CIDG, RF-PF units, and others. They also helped select targets for harassment and interdiction (H&I) and time-on-target (TOT) fires based on air and ground reconnaissance and intelligence reports while also identifying targets of opportunity in support of combat operations. Additionally, they had responsibility for protecting friendly villages from friendly artillery fires and identifying villages suspected of being infiltrated by enemy troops and the Viet Cong based on intelligence reports or direct observation.

Artillery rounds (semi-fixed) – In semi-fixed rounds of artillery ammunition, the projectile (that contains the type of ordnance to be delivered), the cartridge case, powder bags, and fuzes are transported to the battery in separate containers for inspection and temporary storage near each gun. When a fire mission is called, the rounds are individually assembled which allows for the selection of the type of projectile to be delivered, the size of the propellant charge (which is a key determining factors as to the intended range of the shot), and the type fuse required for the fire mission. The predetermined number of powder bags are counted (and verified), then placed inside the cartridge case. The projectile is placed into the metal cartridge case and the fuze is screwed into the projectile tip. Before firing a time fuze the fuze setting is determined and a cannoneer sets the fuze.

Artillery rounds (separate-loading) – In separate-loading artillery ammunition, a complete round consists of four components: a projectile, the fuze, the propellant (in several combustible cloth bags), and the primer.

ARVN – *Army of the Republic of Vietnam*, formed under the auspices of the Republic of South Vietnam and our allies during the Vietnam War. The soldiers of the Republic of South Vietnam. Colloquially pronounced "Arvin," as would rhyme with "Marvin."

ASP – Abbreviation for Ammunition Supply Point, sometimes called an ammo dump or ammunition storage area.

azimuth – Horizontal direction expressed as an angular distance between a fixed point and the direction of an object.

B

BC Scope – A telescopic instrument having a reticle that combined a binocular and a periscope used for targeting, determining distance, estimating range, and artillery adjustment. Depending on design, they sometimes had widely separated objective lenses to aid in more precise distance estimating. *BC* was originally an abbreviation for *Battery Commander*.

Master Sergeant Billy Stout gazing through the BC Scope from the sandbag parapet atop the 6th Battalion, 29th Artillery TOC during the Battle for Dak To in November 1967. At that time, the parapet and BC scope were manned twenty-four hours a day by the Recon/Survey Section. (Larry Skoglund photo)

Band of Brothers – A term borrowed from Shakespeare's play *King Henry V*, Scene 3 (The English Camp) which depicts the king's famous speech to his men on 25 October 1415 (St. Crispin's Day) during the Hundred Years' War to encourage bravery on the morning of the Battle of Agincourt in Northern France. The scene unfolds as English noblemen and their armies under King Henry V gather before the battle and realize that they are greatly outnumbered. One of King Henry's advisors, Westmorland, says that he wishes the men who stayed at home in England had joined them, but the king disagrees. In his speech, King Henry tells his soldiers that they should be happy so few of them are present because each man who is with him will earn an even greater share of the honor. King Henry goes on to say that he does not want to fight alongside any man who does not wish to join the fight and he tells his soldiers that anyone wanting to leave the field of battle can depart, and he will even give them money for their travel home. But, he says, anyone who stays, and fights will have something special to remember and talk about in his old age – something those who remained at home could never do. He adds that even the common men who fight will become his brother, and that all the men who remained at home will eventually regret that they did not join the fight. Although some people today find the speech long-winded and difficult to understand, King Henry's speech in response to Chamberlain's comment is worth reading and may have special significance to some once-young Americans who served in combat during the Vietnam War.

Baptism by Fire (sometimes baptism of fire) – An old military saying meaning being shot at or fired upon the very first time (and surviving). Some say it happens only once, but others disagree.

Battery (field artillery) – A specified number of artillery pieces and the men who man them. The 6th Battalion, 29th Artillery in Vietnam included four batteries – A (alpha); B (bravo); C (Charlie); and H&SB (Headquarters & Service Battery, often called *Headquarters Battery* or *HSB*). In our unit, each battery consisted of six guns, each manned by a gun section, with one gun held in reserve elsewhere.

Battery Commander – The officer commanding an artillery battery and responsible for all its actions, usually the rank of Captain.

Battle of Dak To (technically the Battle for Dak To) – The largest and most extensive engagements of the Vietnam War to that time, was a series of large and smaller military actions fought during most of November 1967 between the U. S. Army and the NVA in the area in and around the forward base of Dak To in the tri-border area of South Vietnam in the Central Highlands.

beehive (artillery round) – An antipersonnel artillery cannister shell filled with over a thousand flechettes and used for direct fire against the enemy. The name originated from the way the flechettes were packed inside the cannister which looked somewhat like a beehive and not (as often incorrectly reported) because they made a sound like buzzing bees. (also see **flechette**)

Bird Dog (commonly Birddog) – The O-1 Cessna observation airplane (see O-1 and O-2). Also, a term commonly used to describe a FAC pilot (Forward Air Controller) or his aircraft whether a Cessna O-1 or O-2.

boonies – A slang term for being out and away from a base camp or fire base. Abbreviation of "boondocks."

boonie hat – A soft-shelled, flat-topped, two and-a-half inch fully brimmed cotton cloth cap worn in the field by soldiers during the Vietnam War. Some were OD while others were camouflaged in various patterns. They were made for the Army and other military branches under U. S. government contracts and also commonly available through several civilian outlets.

Bronze Campaign Star (sometimes Combat Star or Service Star) – A small, bronze, five-pointed star attached to various combat ribbons, each depicting a specific campaign, battle, or military engagement. When more than four bronze service stars are earned, a single silver service star is attached to represent the fifth bronze campaign star. Bronze campaign stars are sometimes incorrectly confused with the Bronze Star Medal which is an individually earned award.

Bronze Star Medal (Bronze Star) – The Bronze Star Medal is the fourth highest honor earned by US Army soldiers during the Vietnam War and was awarded for either service or valor. When earned for valor, a small letter "V" is attached to the ribbon.

C

C-124 – A turboprop, twin engine transportation aircraft used to move men and material that loaded from the rear by way of a ramp. The C-124 was nicknamed *Caribou.*

C-130 – A turboprop, four-engine transportation aircraft used to move men and materiel that is loaded from the rear by way of a ramp. The C-130 was nicknamed *Hercules.*

C-4 – A powerful explosive based on the explosive RDX with a grey or dirty-white color and the consistency of modeling clay with a smell like motor oil. It is very safe to handle and is detonated only by extreme shock such as produced by a blasting cap. C-4 explodes at a rate of about 26,000 feet per second and was used during the Vietnam War for destroying obstacles, tunnels, buildings, and bunkers. It could also be used for clearing bamboo, some trees and other vegetation for helicopter landing zones. It could be safely burned in small quantities and was sometimes used to start small fires for heating C-rations and water for coffee or cocoa when Hexamine fuel tables were unavailable. C-4 was also the explosive used in Claymore mines.

C-rations (*Meal, Combat, Individual* - MCI) – A meal originally developed during WWII to feed troops in the field with individual meals that could be easily prepared quickly to provide nourishment to keep them going. Later, and into the Korean War and Vietnam War, they were improved to offer more variety and better flavor. Each individual meal provided 1,200 calories and was packaged in a brown paperboard box weighing about 1 ½ pounds and labeled with the meal type. Meals were packed twelve to a cardboard case that was secured with two crisscross strands of round wire, so that cases could be handled roughly without breaking open. During the

1967-'68 timeframe, the types of meals changed somewhat although each box always included a protein-based entrée such as chicken and noodles, beans and franks ("beanies and weenies" -the favorite of may), pork slices with juices, ham and eggs (the least favorite of many), ham and lima beans, beefsteak, and probably others. Canned pound cake, pecan cake roll, canned bread or fruit cake was included in some meals as were crackers, peanut butter (had to stir before spreading), sometimes jelly and sometimes canned cheese spread. The canned peaches were everybody's favorite and used for barter or sometimes as a reward for a good deed. Also included in each box was a brown foil bag containing a small packet of instant coffee, sugar, cream substitute, and salt. A tiny roll of toilet paper, a 2-pack box of Chicklets gum, and a 4-pack of brand-name cigarettes was in each box along with a book of paper matches and a white plastic spoon (wrapped in plastic to keep it clean). Each C-ration case also included four P-38 can openers in brown paper wrappers. When available, Hexamine tablets were used to heat C-rations but other times a small chunk of C-4 was used as fuel and occasionally C-ration boxes were burned as were leaves and sticks depending on location and circumstances.

C-ration stove – A small, field-fabricated stove that was easily fashioned from the mostly shorter empty C-ration cans by punching air holes in the base using a triangle can opener (*church key*) or knife, then lighting some type of fuel in the bottom, then placing the open C-ration meal to be heated on top.

Cannon Cockers ("Gun Bunnies") – We called the men on the guns in the firing batteries "Cannon Cockers" or sometimes "Gun Bunnies" neither of which are derogatory terms. The term Gun Bunny supposedly originated when someone observed an artillery battery that was under attack as the Cannoneers quickly jumped the trails of their pieces and moved smartly around to prepare, load, sight, and eagerly await the command to fire. Cannoneers are to artillery as riflemen are to infantry.

Redleg Cannoneer Louie Campbell, Bravo Battery, 6th Battalion, 29th Artillery standing behind the tube of "The Brute," a 105mm Howitzer – both saw heavy action at Dak To. (Photo by Nick Gillen)

Charlie – Our enemy, a name used interchangeably for Viet Cong, NVA, or anybody working for or with the enemy.

cadre – In the U. S. Army, a group of trained leaders, often officers and NCOs, capable of effectively leading and training others in specific skills.

Camp Enari – Headquarters of the Fourth Infantry Division in Vietnam and named in honor of First Lieutenant Mark N. Enari who was killed in action on 2 December 1966, posthumously earning the Silver Star.

CIA – U. S. Central Intelligence Agency, the primary American clandestine organization.

CIDG (Civilian Irregular Defense Group) – A program devised by the CIA in the early 1960s where U. S. Army Special Forces Green Berets began arming and training Montagnards to defend their villages against Viet Cong insurgency. In 1962 at Buon Enao in Darlac Provence, small A-Teams moved into villages and set up Area Development Centers to train for two weeks while other local Strike Force Teams received even better training and weapons to serve as quick reaction forces to respond to Viet Cong attacks. The program was very successful.

CIDG Soldier in Montagnard village near Dak To in November 1967. (Mike Alesi photo)

Claymore mine (*M18A1 antipersonnel mine*) – A directional, ground mounted, horizontally-firing, anti-personnel mine used for ambushes and perimeter defense, usually command detonated using an electrical hand-plunger device (a "*clicker*") that was connected by wire to a blasting cap inserted into one of the two fuze wells set at a 45-degree angle in the top of the mine casing. Inside the casing was a pound and-a-half of C4 explosive behind 700, 1/8" diameter soft steel balls held in place by epoxy resin that were blasted in the direction of the enemy at a velocity of about 4,000 feet per second. The convex shape of the mine (viewed from the front) propelled the balls in an arc of roughly 60 degrees each side and between 6 to 7 feet high to a lethal range of about 50 yards, but with occasional kills out to about a hundred yards. As a very kind reminder, the designers of the mine had the words "*Front Toward Enemy*" embossed on the convex front of the grey-green case. On the top of the case was a simple sighting mechanism for aiming the mine and on the bottom were two retractable scissor-like metal legs for support that could be stuck into soft earth. Early-on, enemy sappers learned to sneak up and turn the Claymore mines around, then make noises to encourage nervous GIs to detonate them, thus causing friendly casualties. At large base camps and more permanent locations, Claymore mines were occasionally set into a concrete-filled wooden forms or still-wet concrete-filled sandbags with rebar stakes driven through them and into the ground to help prevent tampering.

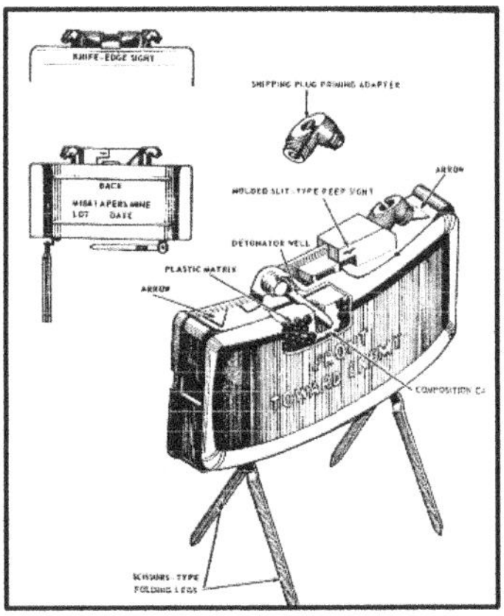

Instructional sketch of M18 Antipersonnel Mine (Claymore Mine) from FM 23-23 (1968)

CO – Commanding Officer. The officer in charge of a military unit.

connex (sometimes CONNEX or "connex box") – A steel container with doors that was developed to ship military equipment during WWII and commonly used during the Vietnam War for equipment shipping, storage, and many other purposes including field TOCs and FDCs. Connex is derived from the words "container, express."

compass – An instrument having a magnetized needle that turns freely on a center axis and points toward magnetic north. Compasses are used (often in conjunction with maps) by artillerymen for the coordination and adjustment of artillery fires and to move predictably from one location to another. A "compass, lensatic," like the one shown in the photo below had a special feature permitting cross-country traversing using an ingenious sighting system and a magnifying glass for viewing the dial for the precise determination of azimuth.

U. S. Army lensatic compass (M-1950) manufactured in December 1966. (Steve Stark photo)

cordite – A smokeless propellant used in some small arms and artillery shells that is classified as a low explosive and formulated from nitroglycerine, gun cotton, and petroleum jelly. When burned, it has a distinct and unique smell.

D

degree (of a circle) – The unit of angular measurement used by infantrymen and others not skilled in the science of artillery where the mil is used instead. There are 360 degrees in a circle which equals exactly 6,400 (six-four-hundred) mils. To obtain an approximation of the conversion of degrees to mils, multiply degrees by 17.7777.

Deuce-and-a-half – The common name for the diesel-powered trucks used extensively in Vietnam for transporting troops and equipment, also sometimes modified as gun platforms. The name "deuce-and-a-half" was a spin-off from the 2.5-ton trucks of WWII (the GMC CCKW). In Vietnam, they were officially designated *Cargo Truck, 1 ½ ton, 6X6 M35.*

Distinguished Service Cross (DSC) – The Distinguished Service Cross is the second highest military decoration that can be earned by a United States Army soldier, always for extraordinary heroism.

DIVARTY – Acronym for Division Artillery.

Dopp kit – A small, usually leather or canvas zippered bag for carrying shaving gear, toothbrush, tooth powder or paste, deodorant, sewing kit, and other men's toiletries and personal items.

duffel bag (commonly spelled duffle) – A heavy canvas (later heavy nylon), olive drab, carrying bag about three feet tall and about twenty inches in diameter with side carrying handle and shoulder strap used for transporting clothing and other gear. The name comes from Duffel, Belgium where a coarse woolen cloth fabric (duffel fabric) was used to make heavy outwear.

E

elevation – The vertical distance, either measured or estimated, above mean sea level and used in land navigation to describe the height of a point. On topographic military maps, elevation is expressed in meters. During the Vietnam War, a hill or a mountain elevation (a "spot elevation") was often used to name a fire support base or hilltop. For example, the well-known battles on 875, 1338, and others were spot elevations on maps of the Dak To area where significant battles were fought, and won, by U. S. Army soldiers and became fire support bases for the 6th Battalion, 29th Artillery and other units.

enlisted man (EM) – During the 1967-'68 timeframe, enlisted men were neither Officers nor Warrant Officers and were classified in one of the following ranks in order of seniority: Private E-1; Private E-2; Private First Class (PFC) E-3; Corporal or Specialist-4, E-4; Sergeant or Specialist-5, E-5; Staff Sergeant or Specialist-6, E-6; Sergeant First Class (SFC) or Specialist-7, E-7; Master Sergeant or First Sergeant, E-8; and Sergeant Major, E-9. Corporals and all Sergeants were classified as Non-Commissioned Officers (often called *Non-Coms*) and sometimes called "hard stripes." Due to differences in training, Specialists' duties typically focused on technical expertise with less personal leadership responsibilities than Corporals and Sergeants although they held the same rank and pay grade.

F

FAC (Forward Air Controller) – The pilot of an observation aircraft (often the fixed-wing, "push-pull" 0-2 propeller aircraft) who located targets from the air and coordinated air strikes by attack aircraft pilots with whom he communicated directly. In our AO in 1967 and '68, FACs were Air Force pilots known collectively as "Bird Dogs" although that term should have been restricted to Army pilots flying the 0-1 propeller aircraft. Like many pilots in Vietnam, FAC pilots had many colorful radio call signs and most performed superbly.

FADAC – Acronym for "Field Artillery Digital Analog Computer" used by an FDC for plotting fires. Capable of incorporating meteorological data into the firing data calculations.

FDC – Abbreviation for Fire Direction Center, the field artillery section manned by soldiers who calculate and establish reliable firing data for a firing battery to help ensure artillery shells and associated munitions get to their intended target.

FM – Abbreviation for Field Manual, the booklets printed by the Army and used for training and instructional purposes.

FO (Forward Observer) – An artilleryman (most often an officer but occasionally enlisted men or NCO) who is trained and skilled in reliably adjusting artillery fires through observation. In our AO during the Vietnam War, there were always shortages of FOs due to their high combat casualty rate.

family jewels – A colloquial term meaning *testicles*, sometimes also referring to other attached body parts located generally between the legs of male members of the U. S. Army and other warriors of the 1960s.

firefight – Military combat by small arms, also applicable to law enforcement officers. Not to be confused with gunfight which is what cowboys sometimes do.

field artillery – The combat branch of the United States Army that employs portable howitzers and guns in support of the infantry and other friendly forces. Field artillery is known as *The King of Battle* because it puts the balls where the queen wants them. The *Queen of Battle* is the infantry.

flechette – Pointed steel nail-like anti-personnel projectiles of various lengths and weighs about 1/16 inch in diameter and having small fins at the rear for stabilization. They were fired point blank from 105mm artillery shells (called *beehive* rounds), from M79 grenade launchers and from aircraft using solid-fuel rockets as the delivery system. The name flechette originated from the French word meaning "little arrow" or "dart." (also see **beehive**)

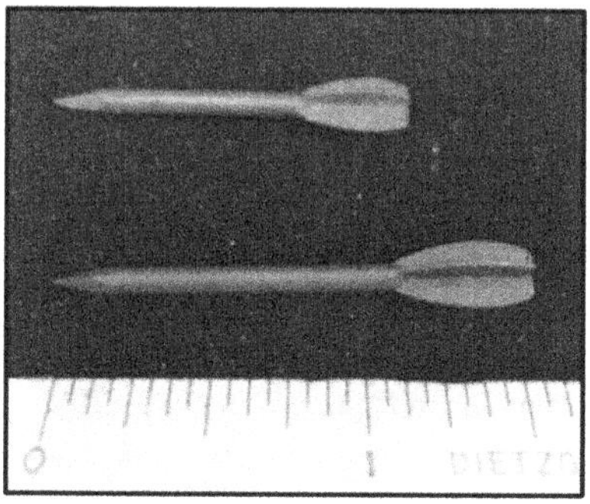

Two sizes of flechettes shown on inch scale – top 7 grain (about 0.016 oz.) and bottom 12 grain (about 0.027 oz.). (Steve Stark photo)

Freedom Bird – The aircraft that took soldiers home from Vietnam.

FSB (Fire Support Base, and often shortened to simply **FB** for Fire Base) – Most FSBs had either number designations or names and often both. Many also had multiple other names or numbers depending on who occupied them and when. Some had very colorful names given by the occupants, often something to do with the location, time of occupancy, or in remembrance of someplace or someone back home. The difference between FSB and an LZ was usually, but not always, that an FSB included artillery.

FUBAR – An acronym originating during WWII sort of like SNAFU. Both terms mean basically the same thing: that everything is thoroughly and completely confused, disordered, damaged, or ruined. It means literally: "Fucked Up Beyond All Recognition" but in polite company, substitute "Fucked" with "Fouled."

full-automatic – A rifle or machinegun designed to fire continuously when the trigger is pressed rearward and continues firing until the trigger is released, when the ammunition supply runs out, or when the weapon jams. Full-automatic is often fired in "bursts," where only a few rounds are fired each time the trigger is pressed briefly and then released.

G

GT Line (Gun-Target Line) – The direct line on a map from the firing battery to the target which is avoided by ground troops in the event of an unintentional long or short round – an artillery round that lands either closer or farther away from the target than was intended.

Geneva Conventions – A set of four international treaties by which wars are supposed to be fought that includes requirements for the humane handling and treatment of prisoners of war, the humane treatment of noncombatants and civilians, care for infirmed and wounded combatants, weapons limitations, and other requirements.
grains (gr.) – a unit of measurement used to quantify the weight of gunpowder and small projectiles such as bullets and flechettes. One ounce is exactly 437.5 grains which means there are 7000 grains in one pound.

Gun Bunny – See "Cannon Cocker."

gunfight – Something cowboys sometimes did on Saturday morning television show when we were kids. Not to be confused with a firefight.

Gunny – In the U. S. Marine Corps, a Gunnery Sergeant holds the rank of E-7, having met specific qualifications in training and leadership.

H

HE (high explosive) – An explosive that detonates at an extremely high rate such as TNT at a rate of approximately 19,000 feet per second. Other explosives, such as black powder, called low explosives burn a much slower rate.

H&I fires – The abbreviation H&I stands for *Harassment and Interdiction* which were pre-planned artillery strikes designed to keep Charlie awake and generally mess up his day by impeding his freedom of movement and destroying his morale, sometimes also wounding or killing him. When planning such fires, extreme caution was exercised to help prevent collateral damage.

I

Infantry – The combat branch of the United States Army that effectively employs ground-based tactics and weaponry to break things and kill and capture bad guys. The U. S. Army infantry is known as the *Queen of Battle.*

in-processing – The U. S. Army process of completing the required paperwork and following unit protocol when a new man joined a unit.

J

jungle boots – Combat boots issued to soldiers in Vietnam that were specially designed for tropical conditions. Although they underwent several design changes during the war, the jungle boots issued to troops of the 6th Battalion, 29th Artillery in 1967 and '68 were constructed with a black leather lower, OD green heavy canvas upper (with ankle protective strip), and special "Panama Pattern" tread designed for muddy conditions. They were also fitted with a protective plate imbedded in the sole for protection from punji stakes and other hazards.

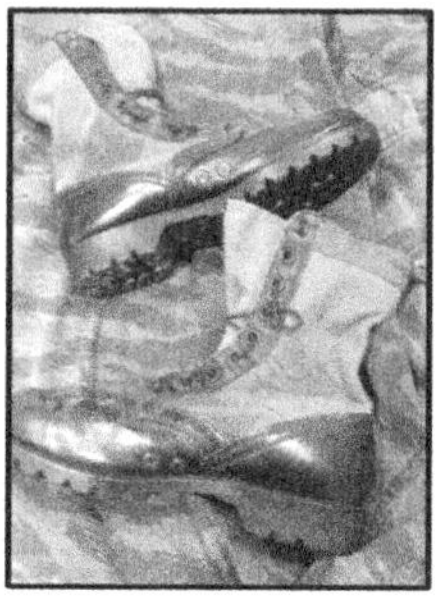

New pair of jungle boots brought home from Dak To by Mike Alesi and gifted Steve Stark during Nashville reunion in 2014, shown on Tiger-Strip pattern fatigue jacket. (Steve Stark photo)

jungle fatigues – The lightweight, lose-fitting, olive drab uniform was designed for hot, humid environments and consisted of trousers and a shirt always worn untucked. The trousers had belt loops (for a web belt), button fly, standard side pockets, rear buttoned pockets, large front-leg buttoned cargo pockets, and cloth blousing ties. The shirts had oversized buttoned breast pockets, large buttoned waist pockets. Like many other types of issued equipment, jungle fatigues underwent several design changes during the Vietnam War.

K

khakis (khaki uniform) – A military uniforms made of khaki material that is sometimes described as a light-colored, yellow-brown cloth made of heavy cotton.

KIA – Killed in Action.

KP – U. S. Army abbreviation for *kitchen police* which means working in a mess hall as usually carried out by enlisted men.

kilometer (colloquial "*click*") – A kilometer is 1,000 meters, 1,093.61 yards, 3,280.83 feet, 0.621371 miles.

L

latrine (sometimes shit house) – The U. S. Army term for a restroom. At the more permanent locations many were elevated wooden outhouses sometimes with multiple holes into which human waste and toilet paper was deposited directly below into 55-gallon drums that had been cut in half and could be removed for burning through hinged doors at the rear.

Burning latrine cans next to very nice four-seater at Dak To (Steve Stark photo)

latrine cans (shit cans) – Usually 55-gallon drums cut in half and used for the collection of human waste, being occasionally dragged using a shit hook from the latrine through a rear-facing hinged door by enlisted men and sanitized by stirring a burning mixture of diesel and gasoline – a duty known as "shit detail."

Loadmaster – The member of an aircrew responsible the safe loading, transport and unloading of aerial cargoes including men and materiel.

low quarter shoes (low quarters) – The U. S. Army's name for Oxford-style, lace-up shoes having a smooth black leather upper, brown leather sole and rubber-cushioned leather heel. Low quarters were worn with khakis or dress uniforms only.

LRRP – Acronym for Long Range Reconnaissance Patrol which were small teams of specially trained U. S. infantry soldiers that were comprised of various numbers of men, often as few as four and sometimes a few more depending on AO and the mission. They sometimes walked out of fire bases but more often were inserted by helicopter to quietly travel the bush to track, locate, monitor, and report on enemy locations, movement and activities. When appropriate, they also conducted raids, attacking and eliminating the enemy. LRRP teams were, by far, responsible for the best intelligence received in the tri-border area and Dak To while we were in country. Reportedly later changed to **LRP** (Long Range Patrol).

LZ – Landing Zone – Usually the name given to a location first occupied by troops arriving by helicopter with most having names as well as numbers, some based on grid coordinates and others having something to do with their location or similarity to someplace or someone back home, often wives or girlfriends or favorite places. The difference between an LZ and an FSB was often, but not always, the presence of artillery.

M

M14 rifle – The M14 rifle fires a 7.62x51mm (30 caliber, called "7.62 NATO") round adopted by the Army in the 1950s to replace the famous M1 Garand that fired a 30.06 round. The 7.62x51 round was an offspring of the civilian .308 Winchester round which, in turn, was based on the .300 Savage hunting cartridge still used by some old guys in Minnesota.

M16 rifle – In the early 1960s the U. S. Army decided that a much lighter, faster-handling rifle was needed to replace the M14. This led to the development of the M16 rifle that was based on the AR15 developed by Eugene Stoner and others at a company called Armalite. The M16 rifle was the primary battle rifle issued to U. S. Army combat soldiers during the Vietnam War beginning in 1965. It was light, gas operated, fed by an external magazine capable of holding 20-rounds of 5.56x45mm ammunition (often loaded only with 18 round) that fired a 55-grain bullet at roughly 2900 feet per second. It produced much lighter recoil compared to earlier U. S. Army main battle rifles such as the M14 and M-1 Garand. The M16 could be fired either semi-automatic or full-automatic depending on the selector switch position on the left side of the receiver. An M16 would completely empty an 18-round magazine in between three and four seconds on continuous automatic fire. During the Vietnam War and thereafter, the M16 underwent several design changes and has since been produced in many variations having separate alphanumeric designations. During the Vietnam War, the M16 was deemed more suited for the often close-in jungle warfare for several reasons including ammunition weight – an M16 and 250 rounds of 5.56mm ammunition in magazines weighed about the same as an M14 and 100 rounds of 7.62x51mm ammo. U. S. Army logic of the time suggested that a soldier carrying a lighter rifle and more bullets was better than a soldier carrying a heavier rifle and fewer bullets.

M60 Machinegun – The M60 machinegun was the primary light machinegun of U. S. forces during the Vietnam War. It fired the same 7.62x54 (7.62 NATO) round that was fired by the M14 rifle.

M79 grenade launcher – A single-shot weapon that fired various types of 40mm projectiles including high explosive rounds and flechettes.

Glen Farasy holding his M79 grenade launcher and shell. (Glen Farasy photo)

Mad Minute – An occasional exercise where we would line up on the perimeter with every piece of man-portable armament we had (M16s, M60s, .30 Carbines, .45 pistols, M79 grenade launchers, LAW rockets) and fire them into the jungle nonstop for one timed minute. Mad minutes were conducted for four reasons: (1.) Eliminate threats hiding in the jungle. (2.) Function-test weapons and clean them afterwards. (3.) Get rid of older ammunition. (4.) Relieve pent-up frustration – it felt good just to raise hell.

MARS – The acronym for Military Affiliate Radio System, a network of ham radio operators established in the mid-1920s to assist with military communications on a local, national, and international basis. Later, it became a network of specially licensed ham radio operators that supported the U. S. Department of Defense. Over the years, MARS volunteers have come primarily from civilian ranks but have also included many active duty military personnel. During the Vietnam War, U. S. military personnel in certain locations were able to communicate stateside during dire emergencies. The MARS acronym changed to Military *Auxiliary* Radio System in 2009 and still operates today through thousands of active participants. The MARS motto was originally "Proudly Serving those who Serve" but later became "Proud, Professional, and Ready."

Medal of Honor – The Medal of Honor is the highest military honor bestowed on an individual soldier that is earned only in combat for valor above and beyond the call of duty.

United States Army Medal of Honor

During the Vietnam War, twelve 4th Infantry Division soldiers earned the Medal of Honor as listed below. The letter "**P**" indicates awarded posthumously (listed by date it was earned).

- Evans, Donald W. Jr., SP/4, Alpha Company, 2nd Bn., 12th Infantry (earned 1/27/67) **P**
- Willett, Louis E., PFC, Charlie Company, 1st Bn., 12th Infantry (earned 2/15/67) **P**
- Smith, Elmelindo R., Sgt., Charlie Company, 2nd Bn., 8th Infantry (earned 2/16/67) **P**
- McNerney, David H., First Sgt., Alpha Company, 1st Bn., 8th Infantry (earned 3/22/67)
- Grandstaff, Bruce A., SFC, Bravo Company, 1st Bn., 8th Infantry (earned 5/18/67) **P**
- Molnar, Frankie Z., Staff Sgt., Bravo Company, 1st Bn., 8th Infantry (earned 5/20/67) **P**
- Bellrichard, Leslie Allen, PFC, Charlie Company, 1st Bn., 8th Infantry (earned 5/20/67) **P**
- Johnson, Dwight H., SP/5, Bravo Company, 1st Bn., 69th Armor (earned 1/15/68)
- Roark, Anund C., Sgt., Charlie Company, 1st Bn., 12th Infantry (earned 5/16/68) **P**
- McDonald, Phill G., PFC, Alpha Company, 1st Bn., 14th Infantry (earned 6/7/68) **P**
- McKibben, Ray, Staff Sgt., Bravo Company, 7th Bn., 17th Cavalry (earned 12/8/68) **P**
- Bennett, Thomas W., CPL, Bravo Company, 1st Bn., 14th Infantry (earned 2/11/69) **P**

mess hall – In the Army, the building, tent, bunker, or area where meals are prepared, and soldiers gather for chow.

meter – A measure of distance used in artillery and other military branches that is equivalent to 1.09361 yards. One kilometer (colloquially "a click") is 1,000 meters which is about 3280.84 feet or about 62/100 of a mile (0.621371 miles).

MIA – Missing in Action.

mil – The unit of angular measurement used by artillerymen for targeting and target adjustment. There are exactly 6,400 (six-four-hundred) mils in a circle and one mil is roughly $1/0.05625^{th}$ of a degree.

Montagnards – Indigenous peoples of the Vietnam Central Highlands who were staunch allies of the U. S. Army. From a French word meaning literally "*mountaineer.*"

mortar – A muzzle-loaded, high-angle, crew served infantry weapon capable of dropping mortar rounds on enemy positions from a high arc. Most mortars can fire many different types of shells including HE, smoke, phosphorous, and illumination. The range of some mortars can be increased by adding booster charges wrapped around the area immediately above the fins. Mortars are sometimes called artillery but are actually an infantry weapon.

Mortar (81mm – A mid-range mortar used by U. S. and allied forces during the Vietnam War with a typical range of approximately 3,200 meters (3,500 yards, 2 miles).

Mortar (82mm) – A mid-range mortar used by the NVA and VC during the Vietnam War that was developed by the USSR in 1938 and has seen extensive use beginning during WWII up to the present time. It had a typical range of approximately 3,000 meters (3,281 yards, 1.9 miles).

Chinese 82mm mortar tail section dug up during a crater analysis near the Dak To airstrip on the afternoon of 12 November 1967. (Steve Stark photo)

Mortar (60mm) – A lightweight, man-portable, mortar used by U. S., NVA, and VC during the Vietnam War, all based on the same basic design similar to 81mm and 82mm mortars. The U. S. and Chicom 60mm mortar with HE projectile had roughly the same maximum range of approximately 1,800 meters (1 mile) and a similar minimum range (with no booster charge) of about 180 meters or 200 yards. The 60mm mortar has sometimes been called "hip pocket artillery" although mortars in the United States Army are typically an infantry deployed weapon.

Mortar (4.2-inch, "Four Deuce") – A heavy mortar typically used on fire bases and at other fixed locations where artillery could not be deployed. It had a minimum range of about 560 meters (560 yards, about 1/3 mile) and a maximum range of approximately 2.7 miles.

MOS – Abbreviation for Military Occupation Specialty. Soldiers successfully completing basic combat training were typically sent to another school (AIT) to learn a specialty. Each specialty had a name and corresponding number.

N

NCO – Abbreviation for Non-Commissioned Officer ("*non-com*") usually holding the rank of sergeant of various grades upward to and including sergeant major but also (in the U. S. Army) including the rank of corporal.

NVA – Abbreviation for North Vietnamese Army.

O

O-1 (L-19 Bird Dog aircraft) – A two-seat, lightweight, high-wing, strut-braced, single-engine, propeller-driven airplane with a tailwheel landing gear designed for the U. S. Army and manufactured by Cessna based on the Cessna 170 design for use in aerial observation and artillery adjustment in a role like the famous Piper L-02 "Grasshopper" flown by artillerymen in WWII. The Bird Dog was fitted with angled side windows to improve ground observation and a clear panel in the rear fuselage that allowed viewing directly rearward. It also had clear panels directly over the cockpit that allowed excellent viewing in many directions while maneuvering at extremely steep angles, often practically inverted, when visibility was most important. A wider door was also installed to allow loading a stretcher or larger equipment for transport. The Bird Dog could take off and land in under 600 feet which made it ideal for the landing strips at Dak To, Dak Pek, and elsewhere in the Central Highlands. It was also flown by U. S. Air Force FAC pilots and was christened "Bird Dog" after a contest held among Cessna employees and the winning entry, submitted by photographer Jack A. Swayze of Wichita, Kansas, was selected by a U.S. Army board. The name was chosen because it fit the role of the Army's new aircraft which was to locate the enemy and orbit overhead as artillery was brought to bear and adjusted to "sniff out and hold" the enemy in a manner like that of a fine hunting bird dog. O-1s were often modified with rocket pods, for firing marking rockets, and with other modifications often depending on need. Also, Bird Dog is often spelled as one word, "Birddog."

O-2 (aircraft, sometimes also called Bird Dog) – A two to four-seat, high-wing, twin-engine (one fore and one aft), dual-propeller (push-pull) airplane designed and manufactured by Cessna for the U. S. Air Force as a replacement for the L-19 (O-1 Bird Dog) using the Cessna 337 Super Skymaster design. Modifications for the military configuration included seating adjustments, see-through door panels, wing-mounted fuel tanks with flame-retardant foam, installing military navigation and communication equipment, and modifications to interior upholstery and removal of other components deemed unnecessary by the Air Force. The first O-2 flew in January 1967 and the USAF took delivery of the first unit just two months later, after which the U. S. Army also began taking delivery. By 1970, 532 O-2s had been built. Some of the O-2s were equipped with loudspeakers and leaflet dispensers for psychological operations use. Also, O-2s used as FACs were equipped with rocket pods and some had the ability to drop ordnance of various types.

Officers – Commissioned soldiers who holds a position of authority over non-commissioned officers and enlisted personnel.

Officers (company-grade) – Officers holding the rank of second lieutenant, first lieutenant, and captain are "company-grade officers."

Officers (field-grade) – Officers holding the rank of major, lieutenant colonel, and colonel are "field-grade officers."

Officers (general grade) – Officers holding the rank of Brigadier General (1 star); Major General (2 stars); Lieutenant General (3 stars); or General (4 stars).

P

P-38 (can opener) – A small metal tool used for opening C-ration cans.

PAVN (People's Army of Vietnam) – another name for the NVA.

PSP – Abbreviation for "Perforated Steel Planking" used for runways and other purposes that is made in strips roughly two feet wide by four feet long which can be linked together through for use as planking for other purposes including covering bunkers over which sandbags could be placed for blast protection.

PSYOPS (sometimes Psy-Ops) – Acronym for Psychological Operations, the process through which efforts are made to control the thinking, and therefore the action, of enemy combatants, civilians, and others.

Poncho Liner – A piece of equipment developed during the Vietnam War that was officially called the "liner, wet weather, poncho" and was made from a thin, white polyester filling that was sandwiched between two layers of camouflaged nylon parachute material that was stitched both crosswise and around its perimeter for strength and formed to fit inside a rubber poncho for comfort, to help trap heat, and keep out the cool night air. It was often used alone as a blanket or shade while on guard duty and became a favorite possession of many U. S. soldiers.

Phoenix Program or **Operation Phoenix** – A CIA program designed to collect intelligence primarily for use against the Viet Cong to help protect American military interests, friendly personnel, and Vietnamese civilians. Their work included kidnappings, interrogations, and assassinations of Viet Cong and those who supported them.

phonetic alphabet – Words used to represent letters of the alphabet to help avoid confusion during radio and telephone messages, sometimes also used in face-to-face conversations for clarity.

A – Alpha
B – Bravo
C – Charlie
D – Delta
E – Echo
F – Foxtrot
G – Golf
H – Hotel
I – India
J – Juliet
K – Kilo
L – Lima
M – Mike
N – November
O – Oscar
P – Papa
Q – Quebec
R – Romeo
S – Sierra
T – Tango
U – Uniform
V – Victor
W – Whisky
X – X-Ray
Y – Yankee
Z – Zulu

piss tube – A simple, field-installed device allowing for quick and easy urination made using empty 105mm, 155mm, or 8-inch powder canister tubes placed at a convenient angle and height into the ground and immediately atop a layer of gravel (when available), or sand or loosened dirt for absorption. In permanent areas, they were sometimes supported in concrete and could be sanitized by occasional burning using a mixture of diesel and gasoline. The tubes were occasionally abandoned and even moved to prevent undue contamination in areas occupied for longer periods of time. Sometimes, a wire screen was placed over the tube, but the reason is still a mystery. Strangely, piss tubes seemed to attract bees, but injuries were typically not reported.

Presidential Unit Citation (PUC) – The Presidential Unit Citation is the highest honor bestowed on a military unit in combat. The 6th Battalion, 29th Artillery earned the PUC during the Battle of Dak To in November 1967 and a smaller contingent of men of the 6th Battalion, 29th Artillery earned a second PUC for combat action on FSB 29 on 25-26 May 1968.

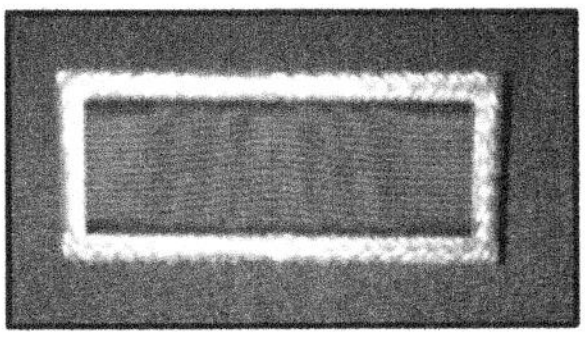

Presidential Unit Citation

Purple Heart – The medal earned by American soldiers wounded in action. Sometimes abbreviated PH.

This Purple heart medal actually represents two separate medals, both earned by 1st Lieutenant Larry Skoglund wounded the first time southwest of Dak To in September 1967 and again in February 1968 just west of the Dak To airstrip. The Oak Leaf Cluster on the ribbon designates his second Purple Heart.

punji stake (sometimes punji stick, punji trap, or just punji); originally *Pungyi* (from a location in Burma) – Any of many types and designs of nasty little mantraps made of wood, bamboo, and sometimes sharpened metal barbs that were placed by Viet Cong and NVA in likely locations, often covered pits and sometimes in rice paddies, where U. S. troops were expected to travel. They were often designed to impale the lower leg and foot and could be difficult to escape without taking time to dig, thus stalling a unit or slowing its momentum. The spikes or sharpened bamboo was sometimes covered with human feces to cause infection. They were intended to wound rather than kill. During WWII in Burma, American soldiers of OSS Detachment 101, including then Colonel William R. Peers, would learn about pungyis from the Kachin people with whom they fought against the Japanese. The Kachins took strips of bamboo about two feet long and cut them into knife-like spears about three-quarters inch wide which were sharpened and then hardened over a fire until they were very strong. They were set into the undergrowth along an ambush trail pointed up so that when the ambush was sprung, the Japanese soldiers would dive into the undergrowth, thus becoming impaled.

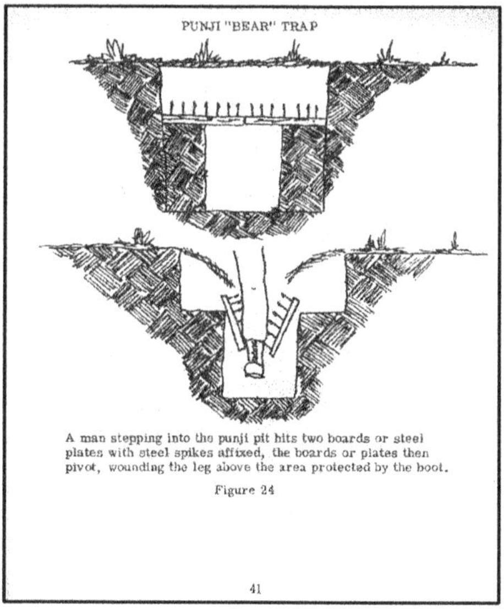
PUNJI "BEAR" TRAP

A man stepping into the punji pit hits two boards or steel plates with steel spikes affixed, the boards or plates then pivot, wounding the leg above the area protected by the boot.

Figure 24

41

Punji Trap – Page 41 of the Handbook for U.S. Forces in Vietnam (issued 1967)

Q

Quad-Fifty (M55 quad .50-cal machinegun) – Four, .50-caliber M2 heavy Browning machineguns on a lightly-armored gun mount, often on a deuce-and-a-half truck ("*gun truck*") or any of a variety of armored vehicles. The four barrels were set in a square pattern and had a maximum range of roughly six and a half kilometers (a little over 4 miles).

RF/PF (sometimes called "*Ruff-Puffs*") – Regional Forces and Popular Forces were lightly armed Vietnamese civilian men and women who fought fight NVA and VC in their villages and hamlets. They were sometimes organized and trained by U. S. Army Special Forces.

Rocket (122mm) – A 4.8-inch diameter unguided rocket of different lengths (all more than 6 feet) with a typical range of between 5 and 20 kilometers (roughly 5 to 12 miles) carrying a 39-pound warhead. Designed to be fired from a tube (the fins were spring-loaded), the NVA often fired them from simple platforms of sticks or embankments that undoubtedly reduced their precision.

This 6-foot long 122mm rocket was captured south of Dak To on "Rocket Ridge" in February 1968 and flown to the First Brigade TOC for inspection. The aluminum carrying case to the right had a metal label written in Vietnamese that was removed to reveal another label with Russian characters. Note the spring-loaded fins and the darker (top) portion of the rocket which was the explosive warhead. (Steve Stark photo)

RTO – Radio Telephone Operator. The soldier who carried a PRC 25 or other small portable FM radios into combat with infantry units for the purpose of coordinating artillery fires.

S

S-1 – Officer in charge of personnel and administration, a Captain's position. In Vietnam he focused on personnel actions and awards dominated by high personnel turnover with soldiers coming into and leaving the battalion on a daily basis.

S-2 – Officer in charge of military intelligence, a Captain's position. In Vietnam his key role was to monitor and interpret intelligence reports from all sources and make recommendations to the Battalion Commander and Operations Officer in formulating operational plans.

S-3 – Officer in charge of operations and training, a Major's position. In Vietnam, he worked with the Battalion Commander and Intelligence Officer and served as the key staff officer in formulating and executing the commander's plan for fire support to the supported infantry battalions.

S-4 – Officer in charge of supply and related matters, a Captain's position. The demands of combat in Vietnam, with combat losses and batteries at distances 40 to 50 miles from base camp (in our case, Camp Enari) offered a constant challenge to the battalion supply section.

SOI (Signal Operating Instructions) – A small code book used for coding and decoding map coordinates and other information that was updated about once every month. When any SOI was lost, all SOIs were considered compromised and new ones were printed and issued.

SNAFU – An acronym originating during WWII sort of like FUBAR. Both terms mean basically the same thing: that everything is thoroughly and completely confused, disordered, damaged, or ruined. It comes from the saying "Situation Normal – All Fucked Up," but in polite company replace "Fucked" with "Fouled."

Sad Sack – A soldier who regularly performs poorly or constantly has problems.

Saint Crispin's Day – A holiday made famous in Shakespeare's play *King Henry V*, Scene 3 (The English Camp) which depicts the king's famous speech to his men on 25 October 1415 (St. Crispin's Day) during the Hundred Years' War to encourage bravery on the morning of the Battle of Agincourt in Northern France where the English, under direct battlefield leadership of King Henry V, defeated a numerically superior French army largely through their prudent application of the famous English long bow (also see Band of Brothers).

Sandbags (sometimes "Bags") – The mascot of the 6th Battalion, 29th Artillery Recon/Survey Section at Dak To (and probably others) in 1967 and part of 1968. She kept us company during the long nights on perimeter duty and lived in our bunker as a faithful companion who had a keen nose for Vietnamese, enemy or otherwise. She had at least one litter of puppies at Dak To.

Our mascot Sandbags on BC Scope duty at Dak To (Steve Stark photo)

Sandbags – In the Central Highlands like most placed in Vietnam during the war, everybody filed sandbags which formed the building blocks of walls of bunkers, often also layered on logs, PSP, or plywood for overhead protection. Stacked as walls about four feet high, they helped protect tents and buildings from shrapnel and offered protection for parked helicopter, fixed-wing aircraft and other equipment. On fire support bases they were used to build open-front chambers for quickly accessing stored artillery projectiles, powder and other components. Properly placed, they offered good protection from small arms fire and shrapnel from mortars, recoilless rifles and rocket-propelled weapons like the RPG and B-40. They offered not only cover but also shade and protection from the rain especially when covered with layers of plastic sheeting. They could be filled and refilled with dirt or sand and used repeatedly when necessary. Even large structures, like the wooden First Brigade Headquarters TOC at Dak To, was completely encased with sandbags. When they were stacked wet or became rain-soaked and then dried, sandbags hardened solid like the adobe bricks used for centuries in the American southwest and elsewhere. We had two kinds of sandbags – cloth and woven plastic and I preferred the cloth ones because they stacked batter and would not slide around on each other as easily as the plastic ones.

Sappers (NVA and VC) – Specially trained and assigned combat troops responsible for breaching perimeters and other obstacles, often using satchel charges once inside the perimeter. They also reportedly set booby-traps, dug tunnels, and planted mines.

Satchel Charge – An offensive weapon using an explosive placed in a satchel, backpack or other carried container sometimes delivered by sappers that is thrown, planted, or otherwise delivered into a bunker, living area, or other enclosure to kill and maim enemy troops. It can be triggered in many ways including by man-lit fuze, mechanical timer, chemical timer, electronic timer, or remotely by wire or radio signal. Satchel charges could be improvised in the field or professionally designed and constructed. Some of the earlies forms were placed into leather satchels to facilitate easier transportation, thus the name.

semi-automatic – A rifle or pistol designed to fire one round each time the trigger is pressed rearward. The trigger can be pressed again and again to fire additional rounds until the magazine is empty.

shit-burning detail (shit detail) – Burning latrine cans – something all enlisted men did from time to time by stirring in lit diesel fuel mixed with gasoline.

Silver Star Medal (*Silver Star*) – The Silver Star Medal was the third highest award earned by U. S. Army soldiers during the Vietnam War. It is awarded only for valor.

Songbird – A derogatory term sometimes used to describe someone who gives information to the enemy rather than conforming to the requirements of the Code of Conduct every American military member is sworn to uphold.

spot elevation – The known elevation of a specific location on a map, often the peak of a hill or mountain or other geographic feature that can be identified from a distance.

Stars and Stripes (sometimes *Stars & Stripes* or *Stars 'n Stripes*) – A newspaper published in cooperation with the U. S. Department of Defense and operating independently and conformant to First Amendment principles.

T

Tet or often TET (Tết in Vietnamese, short for Tết Nguyên Đán) – Translated literally "*Feast of the first morning of the first day*," the most important holiday in Vietnamese culture that celebrates the arrival of spring based on the Vietnamese lunar calendar that usually occurs in January or February.

Tiger Stripes – Special-issue fatigues with a unique camouflage pattern worn primarily by LRRPs and other small specialized units.

TOC – Tactical Operations Center – a place, often a fortified sandbag bunker, from which combat operations were planned and monitored usually for a specific area of operation or unit.

TOT (Time on Target) – Pre-planned strikes where multiple artillery pieces, often at different locations and of different calibers, combine and concentrate the impact of their projectiles on a single target at the same time.

tour of duty – A planned period for soldiers serving in the Republic of South Vietnam, usually twelve consecutive calendar months. The planned tour of duty in 1967 was twelve consecutive months.

tri-border area – The general area of Southeast Asia near the confluence of Laos, Cambodia, and Vietnam.

twin-forty guns (*Dusters*) – Forty-millimeter guns mounted usually on an armored vehicle but sometimes in the back of a deuce-and-a-half truck.

U

UH-1 "HUEY" (helicopter) – The workhorse helicopter of the Vietnam War flown by some of the bravest men ever to walk the earth.

V

VHPA – Vietnam Helicopter Pilots Association.

Valley of Tears – an area of the western Central Highlands south of Dak To where the 4th Infantry Division's 1st Battalion, 8th Infantry and other First Brigade units first fought in Vietnam.

valor – Strength of mind or spirit that enables a person to encounter danger with determination.

Viet Cong (VC) – The Communist-led forces fighting the South Vietnamese government. The political wing was known as the National Liberation Front, and the military was called the People's Liberation Armed Forces. Both the NLF and the PLAF were directed by the People's Revolutionary Party (PRP), the southern branch of the Vietnamese Communist Party, which received direction from Hanoi through COSVN, which was in III Corps on the Cambodian border. After 1968, as negotiations began in Paris, the NLF established the Provisional Revolutionary Government.

Vietnam – A country in Southeast Asia once divided into two separate countries – North Vietnam (controlled by the Communist regime and Ho Chi Minh) and South Vietnam (controlled by democratically-elected representatives). Now recognized internationally as a single country and still under Communist rule.

Vietnam Cross of Gallantry – An award presented by the South Vietnamese government to individuals and units in recognition of their deeds of valor or heroic conduct while engaging the enemy during the Vietnam War. Vietnam veterans who earned this medal must purchase it from civilian sources as it is not provided by the United States Army as are most other medals and awards.

W

WP – White phosphorous (sometimes also "Whiskey-Papa" or "Willy-Pete").

Wasted – Killed.

WIA – Wounded in Action.

X

xin loi (pronounced sin-loy) – a Vietnamese expression meaning "sorry about that."

XO – Executive Officer ("Exec") – In a military unit, the second in command.

Y

yellow – A term sometimes used to describe a cowardly person or acts.

Yellow Brick Road – A sometimes-used code name for the Ho Chi Minh Trail.

Z

Zapped – Shot or otherwise killed.

Zippo – The brand of a liquid-fueled cigarette lighter carried by many GIs, often engraved with places they served or with statements of faith, poetry, or philosophy of life.

After Note:

Our Brother, Patrick J. Reilly, who served with the Recon/Survey Section at Dak To, wrote the following note which we want to share with everyone. Pat served with distinction with the Recon/Survey Section at Dak To and remained in country for two additional tours of duty.

After my initial landing at Cam Ranh Bay, I immediately saw that Air Force accommodations were superior to the Army's – God bless the Air Force. And, yes, Vietnam was different from suburban Nassau County, New York and Montreal, Canada where I grew up.

I had the privilege of serving in the 6th Bn 29th Field Artillery in Vietnam from May 1968 until July 1970, first as an artillery surveyor and later as a forward observer and civil affairs team NCO. Although I had more than one responsibility during my three tours of duty, I always thought of myself as an artillery surveyor first.

I was barely 18 when I arrived at Dak To. Most of the guys in our section were between 20 and 25 years old. And, believe me, when you are only eighteen, being older than twenty is a big deal. Even so, I was immediately accepted into their tightknit group and treated as an equal which I will always remember.

The artillery survey men I served with were intelligent, disciplined and thoughtful – in short, they were good soldiers and superior individuals and they all did their job with very little supervision.

Michael Alesi, Al Roden, Harold Massey, Ted Dieumegard, Tim Tyler, Glen Farasy, Roger Kvasager, Rob Homyak, and Steve Stark were (and still are) true gentlemen.

There is a heaven and I know that Sergeant Leach and Sergeant Price are there, and joining them are Sydney Hegener, Gary Herder and others (forgive me if I forgot any names).

Most of us had descriptive nicknames, and mine was Leper which I brought from home while most others earned theirs in-country and in many different ways.

I liked the Vietnamese and the Montagnards. No doubt they were put in a difficult situation. They did the best they could.

At Dak To our hooch was on the perimeter and upon my arrival I thought the survey section must have done something to aggravate someone up the chain of command to earn that honor.

I missed the battle of Dak To but still there was occasional enemy fire and the knowledge that an attack could occur at any time. But the beer was usually cold, and the cigarettes were free. Of course, we had to "survey-in" 105's when required, drive in convoys, lay concertina wire, fill sandbags and – because we were survey – do all the construction in camp. We all listened to the music: The Rolling Stones, Beatles, Doors etc. Because I was a wrestler, I had everyone doing pushups. Al Roden worked up to substantially over 50. They kept us busy because idle hands do no one any good.

I was promoted to Spec. 5 during my first tour and extended twice for two six-month tours. My first extension was as a forward observer. It was an honor to serve in an officer's position as an E-5. Surveyors knew how to read maps, use a compass and were mathematically competent. I did my job well.

Because I hated the horse size malaria pills, I did not take them and paid the price. While on leave in Canada, my home at the time, I fell ill and was treated at Lakeshore Hospital, but the medical people were unaware that I had malaria. I was finally treated after enduring two weeks of 105 to 106-degree temperatures. That was no way to spend the 30-day leave I was given for extending.

My last six months I served as a civil affairs team chief out of Camp Radcliffe that was decent duty. Like I said, the Vietnamese and Montagnards are good people.

I have two bad memories primarily from my Dak To service. After the U. S. withdrawal, I am certain that Troy, our Montagnard interpreter, was probably executed and our puppy, Sandbags, was probably abandoned.

Every two years the boys from Dak To reunite. That, my friends, is true friendship. God bless all of you.

Pat Reilly (AKA Leper)

Made in the USA
Monee, IL
04 November 2026

67385495R00144